Fourier Analysis

Edited by Paul F. Kisak

Contents

Chapter 1

Fourier analysis

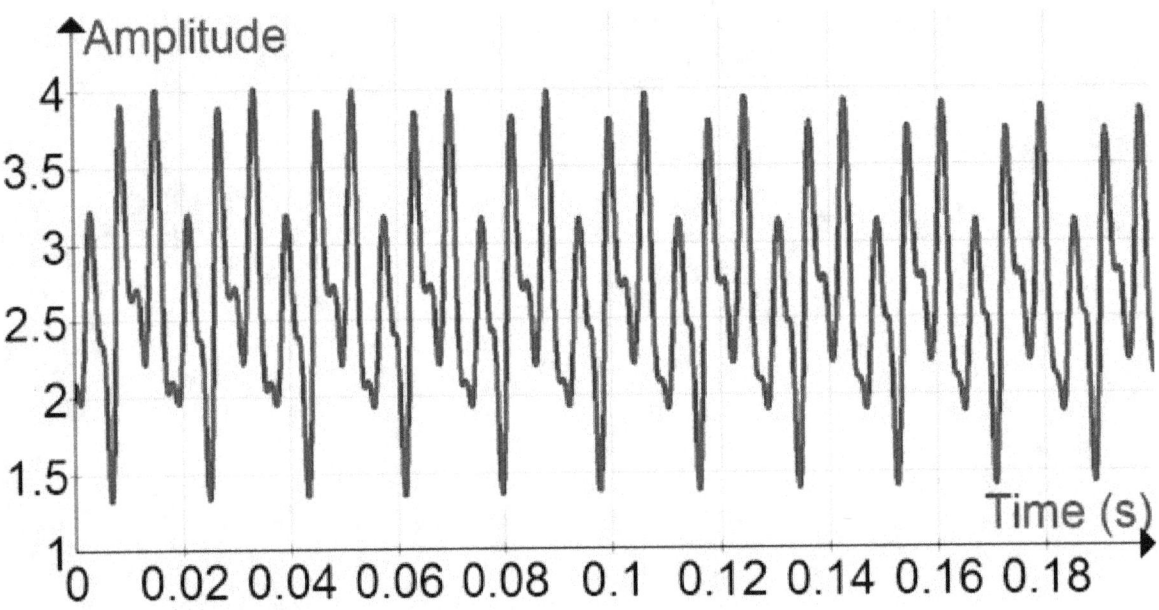

Bass guitar time signal of open string A note (55 Hz).

In mathematics, **Fourier analysis** (English pronunciation: /ˈfɔərieɪ/) is the study of the way general functions may be represented or approximated by sums of simpler trigonometric functions. Fourier analysis grew from the study of Fourier series, and is named after Joseph Fourier, who showed that representing a function as a sum of trigonometric functions greatly simplifies the study of heat transfer.

Today, the subject of Fourier analysis encompasses a vast spectrum of mathematics. In the sciences and engineering, the process of decomposing a function into oscillatory components is often called Fourier analysis, while the operation of rebuilding the function from these pieces is known as **Fourier synthesis**. For example, determining what component frequencies are present in a musical note would involve computing the Fourier transform of a sampled musical note. One could then re-synthesize the same sound by including the frequency components as revealed in the Fourier analysis. In mathematics, the term *Fourier analysis* often refers to the study of both operations.

The decomposition process itself is called a Fourier transformation. Its output, the Fourier transform, is often given a more specific name, which depends upon the domain and other properties of the function being transformed. Moreover, the original concept of Fourier analysis has been extended over time to apply to more and more abstract and general situations, and the general field is often known as harmonic analysis. Each transform used for analysis (see list of Fourier-related transforms) has a corresponding inverse transform that can be used for synthesis.

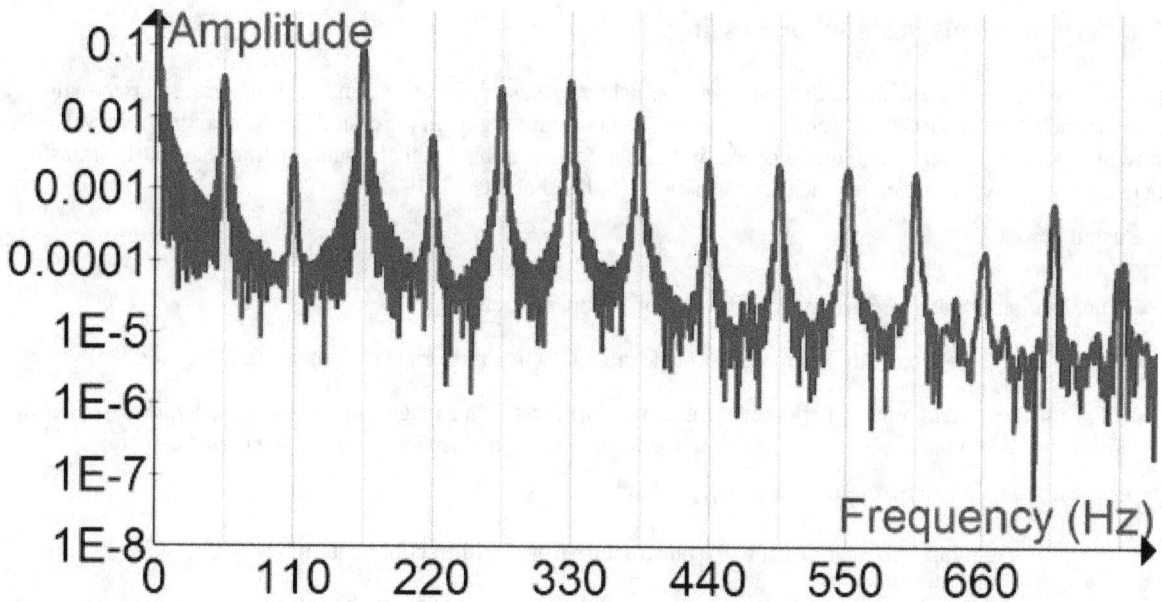

Fourier transform of bass guitar time signal of open string A note (55 Hz), computed with https://sourceforge.net/projects/ amoreaccuratefouriertransform/. Fourier analysis reveals the oscillatory components of signals and functions.

1.1   Applications

Fourier analysis has many scientific applications – in physics, partial differential equations, number theory, combinatorics, signal processing, imaging, probability theory, statistics, option pricing, cryptography, numerical analysis, acoustics, oceanography, sonar, optics, diffraction, geometry, protein structure analysis, and other areas.

This wide applicability stems from many useful properties of the transforms:

- The transforms are linear operators and, with proper normalization, are unitary as well (a property known as Parseval's theorem or, more generally, as the Plancherel theorem, and most generally via Pontryagin duality) (Rudin 1990).

- The transforms are usually invertible.

- The exponential functions are eigenfunctions of differentiation, which means that this representation transforms linear differential equations with constant coefficients into ordinary algebraic ones (Evans 1998). Therefore, the behavior of a linear time-invariant system can be analyzed at each frequency independently.

- By the convolution theorem, Fourier transforms turn the complicated convolution operation into simple multiplication, which means that they provide an efficient way to compute convolution-based operations such as polynomial multiplication and multiplying large numbers (Knuth 1997).

- The discrete version of the Fourier transform (see below) can be evaluated quickly on computers using Fast Fourier Transform (FFT) algorithms. (Conte & de Boor 1980)

Fourier transformation is also useful as a compact representation of a signal. For example, JPEG compression uses a variant of the Fourier transformation (discrete cosine transform) of small square pieces of a digital image. The Fourier components of each square are rounded to lower arithmetic precision, and weak components are eliminated entirely, so that the remaining components can be stored very compactly. In image reconstruction, each image square is reassembled from the preserved approximate Fourier-transformed components, which are then inverse-transformed to produce an approximation of the original image.

1.1.1 Applications in signal processing

When processing signals, such as audio, radio waves, light waves, seismic waves, and even images, Fourier analysis can isolate individual components of a compound waveform, concentrating them for easier detection and/or removal. A large family of signal processing techniques consist of Fourier-transforming a signal, manipulating the Fourier-transformed data in a simple way, and reversing the transformation. (Rabiner and Gold, 1975)

Some examples include:

- Equalization of audio recordings with a series of bandpass filters;

- Digital radio reception with no superheterodyne circuit, as in a modern cell phone or radio scanner;

- Image processing to remove periodic or anisotropic artifacts such as jaggies from interlaced video, stripe artifacts from strip aerial photography, or wave patterns from radio frequency interference in a digital camera;

- Cross correlation of similar images for co-alignment;

- X-ray crystallography to reconstruct a crystal structure from its diffraction pattern;

- Fourier transform ion cyclotron resonance mass spectrometry to determine the mass of ions from the frequency of cyclotron motion in a magnetic field.

- Many other forms of spectroscopy also rely upon Fourier Transforms to determine the three-dimensional structure and/or identity of the sample being analyzed, including Infrared and Nuclear Magnetic Resonance spectroscopies.

- Generation of sound spectrograms used to analyze sounds.

- Passive sonar used to classify targets based on machinery noise.

1.2 Variants of Fourier analysis

1.2.1 (Continuous) Fourier transform

Main article: Fourier transform

Most often, the unqualified term **Fourier transform** refers to the transform of functions of a continuous real argument, and it produces a continuous function of frequency, known as a *frequency distribution*. One function is transformed into another, and the operation is reversible. When the domain of the input (initial) function is time (t), and the domain of the output (final) function is ordinary frequency, the transform of function $s(t)$ at frequency f is given by the complex number:

$$S(f) = \int_{-\infty}^{\infty} s(t) \cdot e^{-i2\pi ft} dt.$$

Evaluating this quantity for all values of f produces the *frequency-domain* function. Then $s(t)$ can be represented as a recombination of complex exponentials of all possible frequencies:

$$s(t) = \int_{-\infty}^{\infty} S(f) \cdot e^{i2\pi ft} df,$$

which is the inverse transform formula. The complex number, $S(f)$, conveys both amplitude and phase of frequency f.

See Fourier transform for much more information, including:

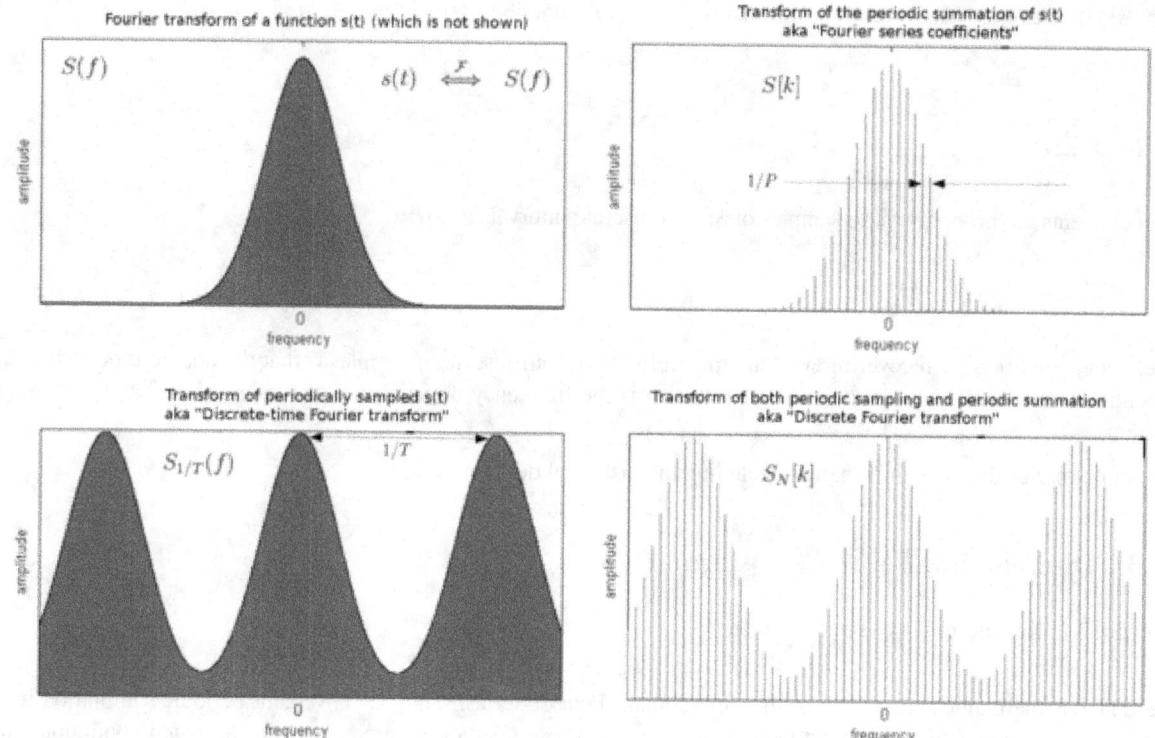

A Fourier transform and 3 variations caused by periodic sampling (at interval T) and/or periodic summation (at interval P) of the underlying time-domain function. The relative computational ease of the DFT sequence and the insight it gives into S(f) make it a popular analysis tool.

- conventions for amplitude normalization and frequency scaling/units

- transform properties

- tabulated transforms of specific functions

- an extension/generalization for functions of multiple dimensions, such as images.

1.2.2   Fourier series

Main article: Fourier series

The Fourier transform of a periodic function, $sP(t)$, with period P, becomes a Dirac comb function, modulated by a sequence of complex coefficients:

$$S[k] = \frac{1}{P} \int_P s_P(t) \cdot e^{-i2\pi \frac{k}{P} t} \, dt \text{ for all integer values of } k,$$

and where $\int_P$ is the integral over any interval of length P.

The inverse transform, known as **Fourier series**, is a representation of $sP(t)$ in terms of a summation of a potentially infinite number of harmonically related sinusoids or complex exponential functions, each with an amplitude and phase specified by one of the coefficients:

$$s_P(t) = \sum_{k=-\infty}^{\infty} S[k] \cdot e^{i2\pi \frac{k}{P} t} \quad \overset{\mathcal{F}}{\Longleftrightarrow} \quad \sum_{k=-\infty}^{+\infty} S[k] \, \delta \left(f - \frac{k}{P} \right).$$

When $sP(t)$, is expressed as a periodic summation of another function, $s(t)$:

$$s_P(t) \overset{\text{def}}{=} \sum_{k=-\infty}^{\infty} s(t - kP),$$

the coefficients are proportional to samples of $S(f)$ at discrete intervals of **1/P**:

$$S[k] = \tfrac{1}{P} \cdot S\left(\tfrac{k}{P}\right). \quad \text{[note 1]}$$

A sufficient condition for recovering $s(t)$ (and therefore $S(f)$) from just these samples is that the non-zero portion of $s(t)$ be confined to a known interval of duration P, which is the frequency domain dual of the Nyquist–Shannon sampling theorem.

See Fourier series for more information, including the historical development.

1.2.3 Discrete-time Fourier transform (DTFT)

Main article: Discrete-time Fourier transform

The DTFT is the mathematical dual of the time-domain Fourier series. Thus, a convergent periodic summation in the frequency domain can be represented by a Fourier series, whose coefficients are samples of a related continuous time function:

$$S_{1/T}(f) \overset{\text{def}}{=} \underbrace{\sum_{k=-\infty}^{\infty} S\left(f - \frac{k}{T}\right)}_{\text{formula summation Poisson}} \equiv \overbrace{\sum_{n=-\infty}^{\infty} s[n] \cdot e^{-i2\pi fnT}}^{\text{(DTFT) series Fourier}} = \mathcal{F}\left\{ \sum_{n=-\infty}^{\infty} s[n]\,\delta(t - nT) \right\},$$

which is known as the DTFT. Thus the **DTFT** of the $s[n]$ sequence is also the **Fourier transform** of the modulated Dirac comb function.[note 2]

The Fourier series coefficients (and inverse transform), are defined by:

$$s[n] \overset{\text{def}}{=} T \int_{1/T} S_{1/T}(f) \cdot e^{i2\pi fnT} df = T \underbrace{\int_{-\infty}^{\infty} S(f) \cdot e^{i2\pi fnT} df}_{\overset{\text{def}}{=}\, s(nT)}$$

Parameter T corresponds to the sampling interval, and this Fourier series can now be recognized as a form of the Poisson summation formula. Thus we have the important result that when a discrete data sequence, $s[n]$, is proportional to samples of an underlying continuous function, $s(t)$, one can observe a periodic summation of the continuous Fourier transform, $S(f)$. That is a cornerstone in the foundation of digital signal processing. Furthermore, under certain idealized conditions one can theoretically recover $S(f)$ and $s(t)$ exactly. A sufficient condition for perfect recovery is that the non-zero portion of $S(f)$ be confined to a known frequency interval of width $1/T$. When that interval is $[-0.5/T, 0.5/T]$, the applicable reconstruction formula is the Whittaker–Shannon interpolation formula.

Another reason to be interested in $S1/T(f)$ is that it often provides insight into the amount of aliasing caused by the sampling process.

Applications of the DTFT are not limited to sampled functions. See Discrete-time Fourier transform for more information on this and other topics, including:

- normalized frequency units

- windowing (finite-length sequences)

- transform properties

- tabulated transforms of specific functions

1.2.4   Discrete Fourier transform (DFT)

Main article: Discrete Fourier transform

The DTFT of a periodic sequence, $sN[n]$, with period N, becomes another Dirac comb function, modulated by the coefficients of a **Fourier series**. And the integral formula for the coefficients simplifies to a summation (see DTFT/Periodic data):

$$S_N[k] = \frac{1}{NT} \underbrace{\sum_N s_N[n] \cdot e^{-i2\pi \frac{k}{N}n}}_{S_k} \text{, where } \textstyle\sum_N \text{ is the sum over any n-sequence of length } \mathbf{N}.$$

The Sk sequence is what's customarily known as the **DFT** of sN. It is also N-periodic, so it is never necessary to compute more than N coefficients. In terms of Sk, the inverse transform is given by:

$$s_N[n] = \frac{1}{N} \textstyle\sum_N S_k \cdot e^{i2\pi \frac{n}{N}k} \text{, where } \textstyle\sum_N \text{ is the sum over any k-sequence of length } \mathbf{N}.$$

When $sN[n]$ is expressed as a periodic summation of another function: $s_N[n] \overset{\text{def}}{=} \sum_{k=-\infty}^{\infty} s[n - kN]$, and $s[n] \overset{\text{def}}{=} T \cdot s(nT)$,

the coefficients are equivalent to samples of $S_{1/T}(f)$ at discrete intervals of $\mathbf{1/P = 1/NT}$: $S_k = S_{1/T}(k/P)$. (see DTFT/Sampling the DTFT)

Conversely, when one wants to compute an arbitrary number (N) of discrete samples of one cycle of a continuous DTFT, $S_{1/T}(f)$, it can be done by computing the relatively simple DFT of $sN[n]$, as defined above. In most cases, N is chosen equal to the length of non-zero portion of $s[n]$. Increasing N, known as *zero-padding* or *interpolation*, results in more closely spaced samples of one cycle of $S1/T(f)$. Decreasing N, causes overlap (adding) in the time-domain (analogous to aliasing), which corresponds to decimation in the frequency domain. (see Sampling the DTFT) In most cases of practical interest, the $s[n]$ sequence represents a longer sequence that was truncated by the application of a finite-length window function or FIR filter array.

The DFT can be computed using a fast Fourier transform (FFT) algorithm, which makes it a practical and important transformation on computers.

See Discrete Fourier transform for much more information, including:

- transform properties

- applications

- tabulated transforms of specific functions

1.2.5   Summary

For periodic functions, both the Fourier transform and the DTFT comprise only a discrete set of frequency components (Fourier series), and the transforms diverge at those frequencies. One common practice (not discussed above) is to handle that divergence via Dirac delta and Dirac comb functions. But the same spectral information can be discerned from just one cycle of the periodic function, since all the other cycles are identical. Similarly, finite-duration functions can be represented as a Fourier series, with no actual loss of information except that the periodicity of the inverse transform is a mere artifact. We also note that none of the formulas here require the duration of s to be limited to the period, **P** or **N**. But that is a common situation, in practice.

In the table below, associating the $\frac{1}{T}$ scale factor with function $S_{1/T}(f)$ results in some notational simplification without loss of generality.

1.2.6 Fourier transforms on arbitrary locally compact abelian topological groups

The Fourier variants can also be generalized to Fourier transforms on arbitrary locally compact abelian topological groups, which are studied in harmonic analysis; there, the Fourier transform takes functions on a group to functions on the dual group. This treatment also allows a general formulation of the convolution theorem, which relates Fourier transforms and convolutions. See also the Pontryagin duality for the generalized underpinnings of the Fourier transform.

1.2.7 Time–frequency transforms

For more details on this topic, see Time–frequency analysis.

In signal processing terms, a function (of time) is a representation of a signal with perfect *time resolution,* but no frequency information, while the Fourier transform has perfect *frequency resolution,* but no time information.

As alternatives to the Fourier transform, in time–frequency analysis, one uses time–frequency transforms to represent signals in a form that has some time information and some frequency information – by the uncertainty principle, there is a trade-off between these. These can be generalizations of the Fourier transform, such as the short-time Fourier transform, the Gabor transform or fractional Fourier transform (FRFT), or can use different functions to represent signals, as in wavelet transforms and chirplet transforms, with the wavelet analog of the (continuous) Fourier transform being the continuous wavelet transform.

1.3 History

See also: Fourier series § Historical development

A primitive form of harmonic series dates back to ancient Babylonian mathematics, where they were used to compute ephemerides (tables of astronomical positions).[1] The classical Greek concepts of deferent and epicycle in the Ptolemaic system of astronomy were related to Fourier series (see Deferent and epicycle: Mathematical formalism).

In modern times, variants of the discrete Fourier transform were used by Alexis Clairaut in 1754 to compute an orbit,[2] which has been described as the first formula for the DFT,[3] and in 1759 by Joseph Louis Lagrange, in computing the coefficients of a trigonometric series for a vibrating string.[4] Technically, Clairaut's work was a cosine-only series (a form of discrete cosine transform), while Lagrange's work was a sine-only series (a form of discrete sine transform); a true cosine+sine DFT was used by Gauss in 1805 for trigonometric interpolation of asteroid orbits.[5] Euler and Lagrange both discretized the vibrating string problem, using what would today be called samples.[4]

An early modern development toward Fourier analysis was the 1770 paper *Réflexions sur la résolution algébrique des équations* by Lagrange, which in the method of Lagrange resolvents used a complex Fourier decomposition to study the solution of a cubic:[6] Lagrange transformed the roots x_1, x_2, x_3 into the resolvents:

$$r_1 = x_1 + x_2 + x_3$$
$$r_2 = x_1 + \zeta x_2 + \zeta^2 x_3$$
$$r_3 = x_1 + \zeta^2 x_2 + \zeta x_3$$

where ζ is a cubic root of unity, which is the DFT of order 3.

A number of authors, notably Jean le Rond d'Alembert, and Carl Friedrich Gauss used trigonometric series to study the heat equation, but the breakthrough development was the 1807 paper *Mémoire sur la propagation de la chaleur dans les corps solides* by Joseph Fourier, whose crucial insight was to model *all* functions by trigonometric series, introducing the Fourier series.

Historians are divided as to how much to credit Lagrange and others for the development of Fourier theory: Daniel Bernoulli and Leonhard Euler had introduced trigonometric representations of functions,[3] and Lagrange had given the Fourier series solution to the wave equation,[3] so Fourier's contribution was mainly the bold claim that an arbitrary function could be represented by a Fourier series.[3]

The subsequent development of the field is known as harmonic analysis, and is also an early instance of representation theory.

The first fast Fourier transform (FFT) algorithm for the DFT was discovered around 1805 by Carl Friedrich Gauss when interpolating measurements of the orbit of the asteroids Juno and Pallas, although that particular FFT algorithm is more often attributed to its modern rediscoverers Cooley and Tukey.[5][7]

1.4   Interpretation in terms of time and frequency

In signal processing, the Fourier transform often takes a time series or a function of continuous time, and maps it into a frequency spectrum. That is, it takes a function from the time domain into the frequency domain; it is a decomposition of a function into sinusoids of different frequencies; in the case of a Fourier series or discrete Fourier transform, the sinusoids are harmonics of the fundamental frequency of the function being analyzed.

When the function f is a function of time and represents a physical signal, the transform has a standard interpretation as the frequency spectrum of the signal. The magnitude of the resulting complex-valued function F at frequency ω represents the amplitude of a frequency component whose initial phase is given by the phase of F.

Fourier transforms are not limited to functions of time, and temporal frequencies. They can equally be applied to analyze *spatial* frequencies, and indeed for nearly any function domain. This justifies their use in such diverse branches as image processing, heat conduction, and automatic control.

1.5   Notes

[1]

$$\int_P \left[\sum_{k=-\infty}^{\infty} s(t-kP) \right] \cdot e^{-i2\pi \frac{k}{P} t} dt = \underbrace{\int_{-\infty}^{\infty} s(t) \cdot e^{-i2\pi \frac{k}{P} t} dt}_{\overset{\text{def}}{=} S(k/P)}$$

[2] We may also note that: $\sum_{n=-\infty}^{+\infty} T\, s(nT)\, \delta(t-nT) = \sum_{n=-\infty}^{+\infty} T\, s(t)\, \delta(t-nT) = s(t) \cdot T \sum_{n=-\infty}^{+\infty} \delta(t-nT)$.
Consequently, a common practice is to model "sampling" as a multiplication by the Dirac comb function, which of course is only "possible" in a purely mathematical sense.

1.6   See also

- Generalized Fourier series

- Fourier-Bessel series

- Fourier-related transforms

- Laplace transform (LT)

- Two-sided Laplace transform

- Mellin transform

- Non-uniform discrete Fourier transform (NDFT)

- Quantum Fourier transform (QFT)

- Number-theoretic transform

- Least-squares spectral analysis

- Basis vectors

- Bispectrum

- Characteristic function (probability theory)

- Orthogonal functions

- Schwartz space

- Spectral density

- Spectral density estimation

- Spectral music

- Wavelet

1.7 Citations

[1] Prestini, Elena (2004), *The evolution of applied harmonic analysis: models of the real world*, Birkhäuser, ISBN 978-0-8176-4125-2, p. 62
Rota, Gian-Carlo; Palombi, Fabrizio (1997), *Indiscrete thoughts*, Birkhäuser, ISBN 978-0-8176-3866-5, p. 11
Neugebauer, Otto (1969) [1957], *The Exact Sciences in Antiquity* (2 ed.), Dover Publications, ISBN 978-0-486-22332-2
Brack-Bernsen, Lis; Brack, Matthias, *Analyzing shell structure from Babylonian and modern times*, arXiv:physics/0310126

[2] Terras, Audrey (1999), *Fourier analysis on finite groups and applications*, Cambridge University Press, ISBN 978-0-521-45718-7, p. 30

[3] Briggs, William L.; Henson, Van Emden (1995), *The DFT : an owner's manual for the discrete Fourier transform*, SIAM, ISBN 978-0-89871-342-8, p. 4

[4] Briggs, William L.; Henson, Van Emden (1995), *The DFT: an owner's manual for the discrete Fourier transform*, SIAM, ISBN 978-0-89871-342-8, p. 2

[5] Heideman, M. T., D. H. Johnson, and C. S. Burrus, "Gauss and the history of the fast Fourier transform," IEEE ASSP Magazine, 1, (4), 14–21 (1984)

[6] Knapp, Anthony W. (2006), *Basic algebra*, Springer, ISBN 978-0-8176-3248-9, p. 501

[7] Terras, Audrey (1999), *Fourier analysis on finite groups and applications*, Cambridge University Press, ISBN 978-0-521-45718-7, p. 31

1.8 References

- Conte, S. D.; de Boor, Carl (1980), *Elementary Numerical Analysis* (Third ed.), New York: McGraw Hill, Inc., ISBN 0-07-066228-2

- Evans, L. (1998), *Partial Differential Equations*, American Mathematical Society, ISBN 3-540-76124-1

- Howell, Kenneth B. (2001). *Principles of Fourier Analysis*, CRC Press. ISBN 978-0-8493-8275-8

- Kamen, E.W., and B.S. Heck. "Fundamentals of Signals and Systems Using the Web and Matlab". ISBN 0-13-017293-6

- Knuth, Donald E. (1997), *The Art of Computer Programming Volume 2: Seminumerical Algorithms* (3rd ed.), Section 4.3.3.C: Discrete Fourier transforms, pg.305: Addison-Wesley Professional, ISBN 0-201-89684-2

- Polyanin, A.D., and A.V. Manzhirov (1998). *Handbook of Integral Equations*, CRC Press, Boca Raton. ISBN 0-8493-2876-4

- Rabiner, Lawrence R., and Bernard Gold. "Theory and application of digital signal processing." Englewood Cliffs, NJ, Prentice-Hall, Inc., 1975. 777 p. 1 (1975).

- Rudin, Walter (1990), *Fourier Analysis on Groups*, Wiley-Interscience, ISBN 0-471-52364-X

- Smith, Steven W. (1999), *The Scientist and Engineer's Guide to Digital Signal Processing* (Second ed.), San Diego, Calif.: California Technical Publishing, ISBN 0-9660176-3-3

- Stein, E.M., and G. Weiss (1971). *Introduction to Fourier Analysis on Euclidean Spaces*. Princeton University Press. ISBN 0-691-08078-X

1.9 External links

- Tables of Integral Transforms at EqWorld: The World of Mathematical Equations.

- An Intuitive Explanation of Fourier Theory by Steven Lehar.

- Lectures on Image Processing: A collection of 18 lectures in pdf format from Vanderbilt University. Lecture 6 is on the 1- and 2-D Fourier Transform. Lectures 7–15 make use of it., by Alan Peters

- Moriarty, Philip; Bowley, Roger (2009). "Σ Summation (and Fourier Analysis)". *Sixty Symbols*. Brady Haran for the University of Nottingham.

Chapter 2

Fourier transform

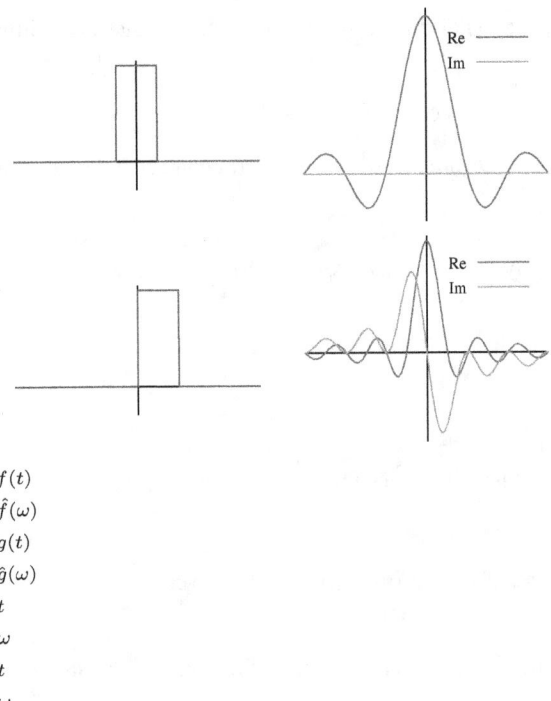

$f(t)$
$\hat{f}(\omega)$
$g(t)$
$\hat{g}(\omega)$
t
ω
t
ω

In the first row is the graph of the unit pulse function $f(t)$ and its Fourier transform $\hat{f}(\omega)$, a function of frequency ω. Translation (that is, delay) in the time domain goes over to complex phase shifts in the frequency domain. In the second row is shown $g(t)$, a delayed unit pulse, beside the real and imaginary parts of the Fourier transform. The Fourier transform decomposes a function into eigenfunctions for the group of translations.

The **Fourier transform** decomposes a function of time (a *signal*) into the frequencies that make it up, similarly to how a musical chord can be expressed as the amplitude (or loudness) of its constituent notes. The Fourier transform of a function of time itself is a complex-valued function of frequency, whose absolute value represents the amount of that frequency present in the original function, and whose complex argument is the phase offset of the basic sinusoid in that frequency. The Fourier transform is called the *frequency domain representation* of the original signal. The term *Fourier transform* refers to both the frequency domain representation and the mathematical operation that associates the frequency domain representation to a function of time. The Fourier transform is not limited to functions of time, but in order to have a unified language, the domain of the original function is commonly referred to as the *time domain*. For many functions of practical interest one can define an operation that reverses this: the *inverse Fourier transformation*, also called *Fourier synthesis*, of a frequency domain representation combines the contributions of all the different frequencies to recover the

11

original function of time.

Linear operations performed in one domain (time or frequency) have corresponding operations in the other domain, which are sometimes easier to perform. The operation of differentiation in the time domain corresponds to multiplication by the frequency,[note 1] so some differential equations are easier to analyze in the frequency domain. Also, convolution in the time domain corresponds to ordinary multiplication in the frequency domain. Concretely, this means that any linear time-invariant system, such as a filter applied to a signal, can be expressed relatively simply as an operation on frequencies.[note 2] After performing the desired operations, transformation of the result can be made back to the time domain. Harmonic analysis is the systematic study of the relationship between the frequency and time domains, including the kinds of functions or operations that are "simpler" in one or the other, and has deep connections to almost all areas of modern mathematics.

Functions that are localized in the time domain have Fourier transforms that are spread out across the frequency domain and vice versa, a phenomenon known as the uncertainty principle. The critical case for this principle is the Gaussian function, of substantial importance in probability theory and statistics as well as in the study of physical phenomena exhibiting normal distribution (e.g., diffusion). The Fourier transform of a Gaussian function is another Gaussian function. Joseph Fourier introduced the transform in his study of heat transfer, where Gaussian functions appear as solutions of the heat equation.

The Fourier transform can be formally defined as an improper Riemann integral, making it an integral transform, although this definition is not suitable for many applications requiring a more sophisticated integration theory.[note 3] For example, many relatively simple applications use the Dirac delta function, which can be treated formally as if it were a function, but the justification requires a mathematically more sophisticated viewpoint.[1] The Fourier transform can also be generalized to functions of several variables on Euclidean space, sending a function of 3-dimensional space to a function of 3-dimensional momentum (or a function of space and time to a function of 4-momentum). This idea makes the spatial Fourier transform very natural in the study of waves, as well as in quantum mechanics, where it is important to be able to represent wave solutions as functions of either space or momentum and sometimes both. In general, functions to which Fourier methods are applicable are complex-valued, and possibly vector-valued.[2] Still further generalization is possible to functions on groups, which, besides the original Fourier transform on $\mathbb{R}$ or $\mathbb{R}^n$ (viewed as groups under addition), notably includes the discrete-time Fourier transform (DTFT, group = $\mathbb{Z}$), the discrete Fourier transform (DFT, group = $\mathbb{Z}$ mod N) and the Fourier series or circular Fourier transform (group = S^1, the unit circle $\approx$ closed finite interval with endpoints identified). The latter is routinely employed to handle periodic functions. The fast Fourier transform (FFT) is an algorithm for computing the DFT.

2.1   Definition

There are several common conventions for defining the Fourier transform $\hat{f}$ of an integrable function $f : \mathbb{R} \to \mathbb{C}$ (Kaiser 1994, p. 29), (Rahman 2011, p. 11). This article will use the following definition:

$$\hat{f}(\xi) = \int_{-\infty}^{\infty} f(x)\, e^{-2\pi i x \xi}\, dx, \text{ for any real number } \xi.$$

When the independent variable x represents *time* (with SI unit of seconds), the transform variable ξ represents frequency (in hertz). Under suitable conditions, f is determined by $\hat{f}$ via the **inverse transform**:

$$f(x) = \int_{-\infty}^{\infty} \hat{f}(\xi)\, e^{2\pi i \xi x}\, d\xi, \text{ for any real number } x.$$

The statement that f can be reconstructed from $\hat{f}$ is known as the Fourier inversion theorem, and was first introduced in Fourier's *Analytical Theory of Heat* (Fourier 1822, p. 525), (Fourier & Freeman 1878, p. 408), although what would be considered a proof by modern standards was not given until much later (Titchmarsh 1948, p. 1). The functions f and $\hat{f}$ often are referred to as a *Fourier integral pair* or *Fourier transform pair* (Rahman 2011, p. 10).

For other common conventions and notations, including using the angular frequency ω instead of the frequency ξ, see Other conventions and Other notations below. The Fourier transform on Euclidean space is treated separately, in which the variable x often represents position and ξ momentum.

2.2 History

Main articles: Fourier analysis § History and Fourier series § History

In 1822, Joseph Fourier showed that some functions could be written as an infinite sum of harmonics.[3]

2.3 Introduction

See also: Fourier analysis
 One motivation for the Fourier transform comes from the study of Fourier series. In the study of Fourier series, com-

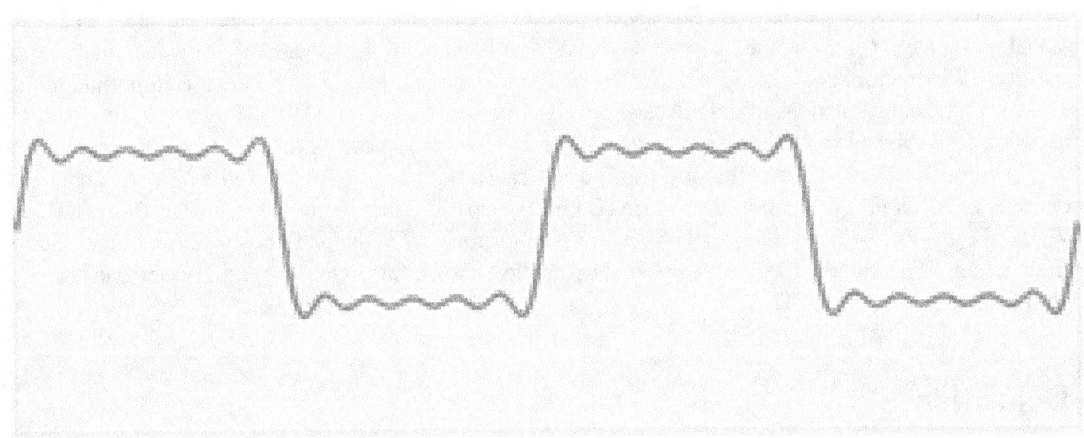

In the first frames of the animation, a function f is resolved into Fourier series: a linear combination of sines and cosines (in blue). The component frequencies of these sines and cosines spread across the frequency spectrum, are represented as peaks in the frequency domain (actually Dirac delta functions, shown in the last frames of the animation). The frequency domain representation of the function, f̂ , is the collection of these peaks at the frequencies that appear in this resolution of the function.

plicated but periodic functions are written as the sum of simple waves mathematically represented by sines and cosines. The Fourier transform is an extension of the Fourier series that results when the period of the represented function is lengthened and allowed to approach infinity (Taneja 2008, p. 192).

Due to the properties of sine and cosine, it is possible to recover the amplitude of each wave in a Fourier series using an integral. In many cases it is desirable to use Euler's formula, which states that $e^{2\pi i\theta} = \cos(2\pi\theta) + i\sin(2\pi\theta)$, to write

Fourier series in terms of the basic waves $e^{2\pi i \theta}$. This has the advantage of simplifying many of the formulas involved, and provides a formulation for Fourier series that more closely resembles the definition followed in this article. Re-writing sines and cosines as complex exponentials makes it necessary for the Fourier coefficients to be complex valued. The usual interpretation of this complex number is that it gives both the amplitude (or size) of the wave present in the function and the phase (or the initial angle) of the wave. These complex exponentials sometimes contain negative "frequencies". If θ is measured in seconds, then the waves $e^{2\pi i \theta}$ and $e^{-2\pi i \theta}$ both complete one cycle per second, but they represent different frequencies in the Fourier transform. Hence, frequency no longer measures the number of cycles per unit time, but is still closely related.

There is a close connection between the definition of Fourier series and the Fourier transform for functions f that are zero outside an interval. For such a function, we can calculate its Fourier series on any interval that includes the points where f is not identically zero. The Fourier transform is also defined for such a function. As we increase the length of the interval on which we calculate the Fourier series, then the Fourier series coefficients begin to look like the Fourier transform and the sum of the Fourier series of f begins to look like the inverse Fourier transform. To explain this more precisely, suppose that T is large enough so that the interval $[-T/2, T/2]$ contains the interval on which f is not identically zero. Then the n-th series coefficient cn is given by:

$$c_n = \frac{1}{T} \int_{-T/2}^{T/2} f(x) \, e^{-2\pi i (n/T)x} \, dx.$$

Comparing this to the definition of the Fourier transform, it follows that $c_n = (1/T)\hat{f}(n/T)$ since $f(x)$ is zero outside $[-T/2, T/2]$. Thus the Fourier coefficients are just the values of the Fourier transform sampled on a grid of width $1/T$, multiplied by the grid width $1/T$.

Under appropriate conditions, the Fourier series of f will equal the function f. In other words, f can be written:

$$f(x) = \sum_{n=-\infty}^{\infty} c_n \, e^{2\pi i (n/T)x} = \sum_{n=-\infty}^{\infty} \hat{f}(\xi_n) \, e^{2\pi i \xi_n x} \Delta\xi,$$

where the last sum is simply the first sum rewritten using the definitions $\xi n = n/T$, and $\Delta\xi = (n+1)/T - n/T = 1/T$.

This second sum is a Riemann sum, and so by letting $T \to \infty$ it will converge to the integral for the inverse Fourier transform given in the definition section. Under suitable conditions this argument may be made precise (Stein & Shakarchi 2003).

In the study of Fourier series the numbers cn could be thought of as the "amount" of the wave present in the Fourier series of f. Similarly, as seen above, the Fourier transform can be thought of as a function that measures how much of each individual frequency is present in our function f, and we can recombine these waves by using an integral (or "continuous sum") to reproduce the original function.

2.4 Example

The following figures provide a visual illustration of how the Fourier transform measures whether a frequency is present in a particular function. The function depicted $f(t) = \cos(6\pi t)\, e^{-\pi t^2}$ oscillates at 3 Hz (if t measures seconds) and tends quickly to 0. (The second factor in this equation is an envelope function that shapes the continuous sinusoid into a short pulse. Its general form is a Gaussian function). This function was specially chosen to have a real Fourier transform that can easily be plotted. The first image contains its graph. In order to calculate $\hat{f}(3)$ we must integrate $e^{-2\pi i(3t)} f(t)$. The second image shows the plot of the real and imaginary parts of this function. The real part of the integrand is almost always positive, because when $f(t)$ is negative, the real part of $e^{-2\pi i(3t)}$ is negative as well. Because they oscillate at the same rate, when $f(t)$ is positive, so is the real part of $e^{-2\pi i(3t)}$. The result is that when you integrate the real part of the integrand you get a relatively large number (in this case 0.5). On the other hand, when you try to measure a frequency that is not present, as in the case when we look at $\hat{f}(5)$, you see that both real and imaginary component of this function vary rapidly between positive and negative values, as plotted in the third image. Therefore, in this case, the integrand oscillates fast enough so that the integral is very small and the value for the fourier transform for that frequency is nearly zero.

The general situation may be a bit more complicated than this, but this in spirit is how the Fourier transform measures how much of an individual frequency is present in a function $f(t)$.

- Original function showing oscillation 3 Hz.

- Real and imaginary parts of integrand for Fourier transform at 3 Hz

- Real and imaginary parts of integrand for Fourier transform at 5 Hz

- Fourier transform with 3 and 5 Hz labeled.

2.5 Properties of the Fourier transform

Here we assume $f(x)$, $g(x)$ and $h(x)$ are *integrable functions*: Lebesgue-measurable on the real line satisfying:

$$\int_{-\infty}^{\infty} |f(x)|\, dx < \infty.$$

We denote the Fourier transforms of these functions by $\hat{f}(\xi)$, $\hat{g}(\xi)$ and $\hat{h}(\xi)$ respectively.

2.5.1 Basic properties

The Fourier transform has the following basic properties: (Pinsky 2002).

Linearity

For any complex numbers a and b, if $h(x) = af(x) + bg(x)$, then $\hat{h}(\xi) = a \cdot \hat{f}(\xi) + b \cdot \hat{g}(\xi)$.

Translation/ Time-Shifting

For any real number x_0, if $h(x) = f(x - x_0)$, then $\hat{h}(\xi) = e^{-i\,2\pi\,x_0\,\xi}\hat{f}(\xi)$.

Modulation/ Frequency shifting

For any real number ξ_0 if $h(x) = e^{i\,2\pi\,x\,\xi_0} f(x)$, then $\hat{h}(\xi) = \hat{f}(\xi - \xi_0)$.

Time Scaling

For a non-zero real number a, if $h(x) = f(ax)$, then $\hat{h}(\xi) = \frac{1}{|a|}\hat{f}\left(\frac{\xi}{a}\right)$. The case $a = -1$ leads to the *time-reversal* property, which states: if $h(x) = f(-x)$, then $\hat{h}(\xi) = \hat{f}(-\xi)$.

Conjugation

If $h(x) = \overline{f(x)}$, then $\hat{h}(\xi) = \overline{\hat{f}(-\xi)}$.

In particular, if f is real, then one has the *reality condition* $\hat{f}(-\xi) = \overline{\hat{f}(\xi)}$, that is, $\hat{f}$ is a Hermitian function.

And if f is purely imaginary, then $\hat{f}(-\xi) = -\overline{\hat{f}(\xi)}$.

Integration

$\xi = 0$

$$\hat{f}(0) = \int_{-\infty}^{\infty} f(x)\, dx.$$

That is, the evaluation of the Fourier transform in the origin ($\xi = 0$) equals the integral of f over all its domain.

2.5.2 Invertibility and periodicity

Further information: Fourier inversion theorem and Fractional Fourier transform

Under suitable conditions on the function f, it can be recovered from its Fourier transform $\hat{f}$. Indeed, denoting the Fourier transform operator by $\mathcal{F}$, so $\mathcal{F}(f) := \hat{f}$, then for suitable functions, applying the Fourier transform twice simply flips the function: $\mathcal{F}^2(f)(x) = f(-x)$, , which can be interpreted as "reversing time". Since reversing time is two-periodic, applying this twice yields $\mathcal{F}^4(f) = f$, so the Fourier transform operator is four-periodic, and similarly the inverse Fourier transform can be obtained by applying the Fourier transform three times: $\mathcal{F}^3(\hat{f}) = f$. In particular the Fourier transform is invertible (under suitable conditions).

More precisely, defining the **parity operator** $\mathcal{P}$ that inverts time, $\mathcal{P}[f]: t \mapsto f(-t)$, :

$$\mathcal{F}^0 = \text{Id}, \qquad \mathcal{F}^1 = \mathcal{F}, \qquad \mathcal{F}^2 = \mathcal{P}, \qquad \mathcal{F}^4 = \text{Id}$$

$$\mathcal{F}^3 = \mathcal{F}^{-1} = \mathcal{P} \circ \mathcal{F} = \mathcal{F} \circ \mathcal{P}$$

These equalities of operators require careful definition of the space of functions in question, defining equality of functions (equality at every point? equality almost everywhere?) and defining equality of operators – that is, defining the topology on the function space and operator space in question. These are not true for all functions, but are true under various conditions, which are the content of the various forms of the Fourier inversion theorem.

This four-fold periodicity of the Fourier transform is similar to a rotation of the plane by $90°$, particularly as the two-fold iteration yields a reversal, and in fact this analogy can be made precise. While the Fourier transform can simply be interpreted as switching the time domain and the frequency domain, with the inverse Fourier transform switching them back, more geometrically it can be interpreted as a rotation by $90°$ in the time–frequency domain (considering time as the x-axis and frequency as the y-axis), and the Fourier transform can be generalized to the fractional Fourier transform, which involves rotations by other angles. This can be further generalized to linear canonical transformations, which can be visualized as the action of the special linear group $\text{SL}_2(\mathbf{R})$ on the time–frequency plane, with the preserved symplectic form corresponding to the uncertainty principle, below. This approach is particularly studied in signal processing, under time–frequency analysis.

2.5.3 Units and Duality

In mathematics, one often does not think of any units as being attached to the two variables t and ξ . But in physical applications, ξ must have inverse units to the units of t . For example, if t is measured in seconds, ξ should be in cycles per second for the formulas here to be valid. If the scale of t is changed and t is measured in units of 2π seconds, then either ξ must be in the so-called "angular frequency", or one must insert some constant scale factor into some of the formulas. If t is measured in units of length, then ξ must be in inverse length, e.g., wavenumbers. That is to say, there are two copies of the real line: one measured in one set of units, where t ranges, and the other in inverse units to the units of t , and which is the range of ξ . So these are two distinct copies of the real line, and cannot be identified with each other. Therefore, the Fourier transform goes from one space of functions to a different space of functions: functions which have a different domain of definition.

In general, ξ must always be taken to be a linear form on the space of t s, which is to say that the second real line is the dual space of the first real line. See the article on linear algebra for a more formal explanation and for more details. This point of view becomes essential in generalisations of the Fourier transform to general symmetry groups, including the case of Fourier series.

That there is no one preferred way (often, one says "no canonical way") to compare the two copies of the real line which are involved in the Fourier transform—fixing the units on one line does not force the scale of the units on the other line—is the reason for the plethora of rival conventions on the definition of the Fourier transform. The various definitions resulting from different choices of units differ by various constants. If the units of t are in seconds but the units of ξ are in angular frequency, then the angular frequency variable is often denoted by one or another Greek letter, for example, $\omega = 2\pi\xi$ is quite common. Thus (writing $\hat{x}_1$ for the alternative definition and $\hat{x}$ for the definition adopted in this article)

$$\hat{x}_1(\omega) = \hat{x}\left(\frac{\omega}{2\pi}\right) = \int_{-\infty}^{\infty} x(t)e^{-i\omega t}\,dt$$

as before, but the corresponding alternative inversion formula would then have to be

$$x(t) = \frac{1}{2\pi}\int_{-\infty}^{\infty} \hat{x}_1(\omega)e^{it\omega}\,d\omega.$$

To have something involving angular frequency but with greater symmetry between the Fourier transform and the inversion formula, one very often sees still another alternative definition of the Fourier transform, with a factor of $\sqrt{2\pi}$, thus

$$\hat{x}_2(\omega) = \frac{1}{\sqrt{2\pi}}\int_{-\infty}^{\infty} x(t)e^{-i\omega t}\,dt,$$

and the corresponding inversion formula then has to be

$$x(t) = \frac{1}{\sqrt{2\pi}}\int_{-\infty}^{\infty} \hat{x}_2(\omega)e^{it\omega}\,d\omega.$$

Furthermore, there is no way to fix which square root of negative one will be meant by the symbol i (it makes no sense to speak of "the positive square root" since only real numbers can be positive, similarly it makes no sense to say "rotation counter-clockwise", because until i is chosen, there is no fixed way to draw the complex plane), and hence one occasionally sees the Fourier transform written with i in the exponent instead of $-i$, and vice versa for the inversion formula, a convention that is equally valid as the one chosen in this article, which is the more usual one.

For example, in probability theory, the characteristic function ϕ of the probability density function f of a random variable X of continuous type is defined without a negative sign in the exponential, and since the units of x are ignored, there is no 2π either:

$$\phi(\lambda) = \int_{-\infty}^{\infty} f(x)e^{i\lambda x}\,dx.$$

(In probability theory, and in mathematical statistics, the use of the Fourier—Stieltjes transform is preferred, because so many random variables are not of continuous type, and do not possess a density function, and one must treat discontinuous distribution functions, i.e., measures which possess "atoms".)

From the higher point of view of group characters, which is much more abstract, all these arbitrary choices disappear, as will be explained in the later section of this article, on the notion of the Fourier transform of a function on an Abelian locally compact group.

2.5.4 Uniform continuity and the Riemann–Lebesgue lemma

The Fourier transform may be defined in some cases for non-integrable functions, but the Fourier transforms of integrable functions have several strong properties.

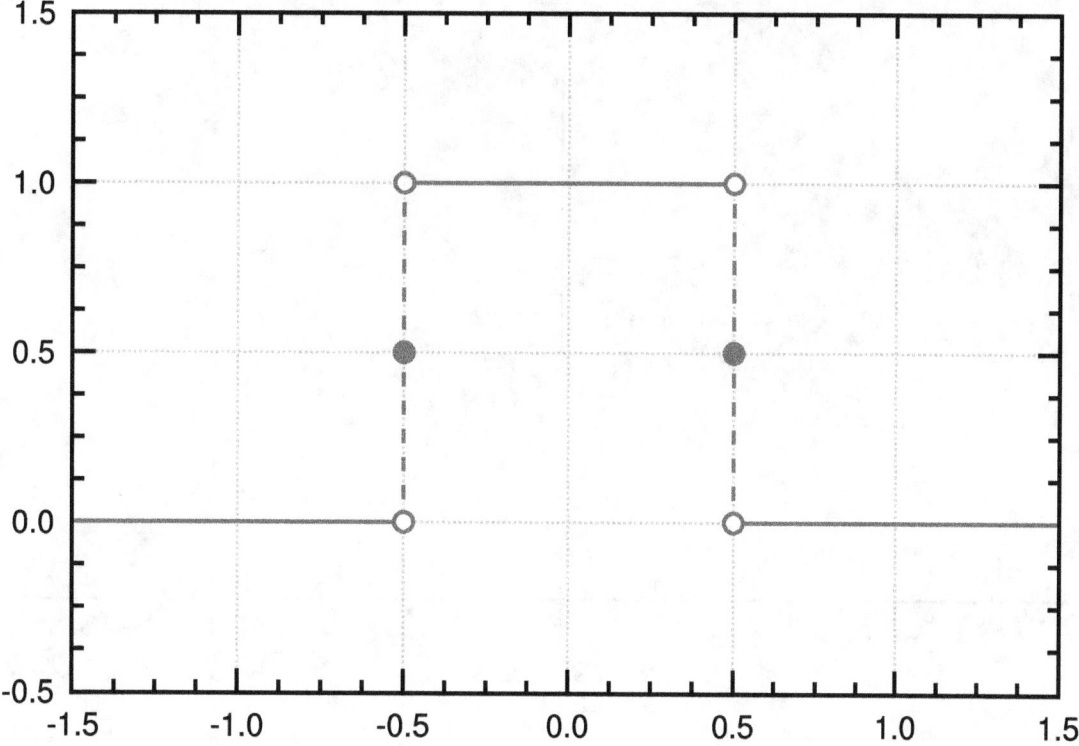

The rectangular function is Lebesgue integrable.

The Fourier transform, $\hat{f}$, of any integrable function f is uniformly continuous and $\|\hat{f}\|_\infty \leq \|f\|_1$ (Katznelson 1976). By the *Riemann–Lebesgue lemma* (Stein & Weiss 1971),

$$\hat{f}(\xi) \to 0 \text{ as } |\xi| \to \infty.$$

However, $\hat{f}$ need not be integrable. For example, the Fourier transform of the rectangular function, which is integrable, is the sinc function, which is not Lebesgue integrable, because its improper integrals behave analogously to the alternating harmonic series, in converging to a sum without being absolutely convergent.

It is not generally possible to write the *inverse transform* as a Lebesgue integral. However, when both f and $\hat{f}$ are integrable, the inverse equality

$$f(x) = \int_{-\infty}^{\infty} \hat{f}(\xi)e^{2i\pi x\xi}\,d\xi$$

holds almost everywhere. That is, the Fourier transform is injective on $L^1(\mathbf{R})$. (But if f is continuous, then equality holds for every x.)

2.5.5 Plancherel theorem and Parseval's theorem

Let $f(x)$ and $g(x)$ be integrable, and let $\hat{f}(\xi)$ and $\hat{g}(\xi)$ be their Fourier transforms. If $f(x)$ and $g(x)$ are also square-integrable, then we have Parseval's Formula (Rudin 1987, p. 187):

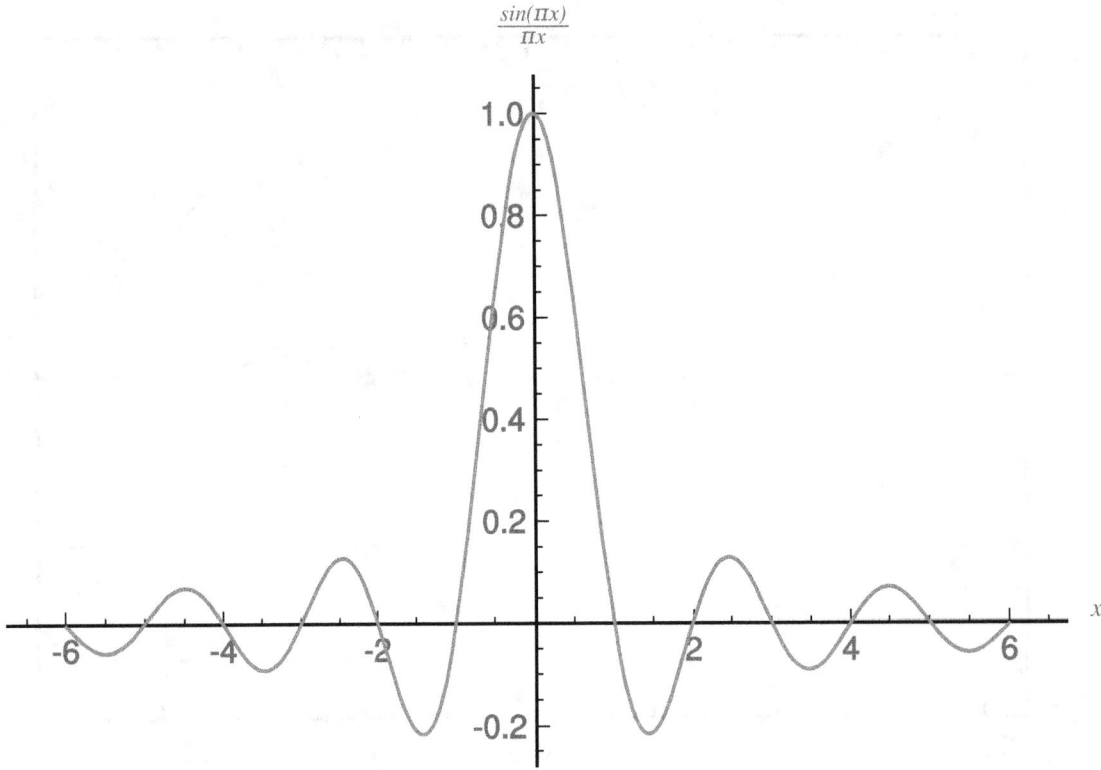

The sinc function, which is the Fourier transform of the rectangular function, is bounded and continuous, but not Lebesgue integrable.

$$\int_{-\infty}^{\infty} f(x)\overline{g(x)}\,\mathrm{d}x = \int_{-\infty}^{\infty} \hat{f}(\xi)\overline{\hat{g}(\xi)}\,d\xi,$$

where the bar denotes complex conjugation.

The Plancherel theorem, which follows from the above, states that (Rudin 1987, p. 186)

$$\int_{-\infty}^{\infty} |f(x)|^2\,dx = \int_{-\infty}^{\infty} \left|\hat{f}(\xi)\right|^2\,d\xi.$$

Plancherel's theorem makes it possible to extend the Fourier transform, by a continuity argument, to a unitary operator on $L^2(\mathbf{R})$. On $L^1(\mathbf{R}) \cap L^2(\mathbf{R})$, this extension agrees with original Fourier transform defined on $L^1(\mathbf{R})$, thus enlarging the domain of the Fourier transform to $L^1(\mathbf{R}) + L^2(\mathbf{R})$ (and consequently to $L^p(\mathbf{R})$ for $1 \le p \le 2$). Plancherel's theorem has the interpretation in the sciences that the Fourier transform preserves the energy of the original quantity. The terminology of these formulas is not quite standardised. Parseval's theorem was proved only for Fourier series, and was first proved by Liapounoff. But Parseval's formula makes sense for the Fourier transform as well, and so even though in the context of the Fourier transform it was proved by Plancherel, it is still often referred to as Parseval's formula, or Parseval's relation, or even Parseval's theorem.

See Pontryagin duality for a general formulation of this concept in the context of locally compact abelian groups.

2.5.6   Poisson summation formula

Main article: Poisson summation formula

The Poisson summation formula (PSF) is an equation that relates the Fourier series coefficients of the periodic summation of a function to values of the function's continuous Fourier transform. The Poisson summation formula says that for sufficiently regular functions f,

$$\sum_n \hat{f}(n) = \sum_n f(n).$$

It has a variety of useful forms that are derived from the basic one by application of the Fourier transform's scaling and time-shifting properties. The formula has applications in engineering, physics, and number theory. The frequency-domain dual of the standard Poisson summation formula is also called the discrete-time Fourier transform.

Poisson summation is generally associated with the physics of periodic media, such as heat conduction on a circle. The fundamental solution of the heat equation on a circle is called a theta function. It is used in number theory to prove the transformation properties of theta functions, which turn out to be a type of modular form, and it is connected more generally to the theory of automorphic forms where it appears on one side of the Selberg trace formula.

2.5.7   Differentiation

Suppose $f(x)$ is a differentiable function, and both f and its derivative f' are integrable. Then the Fourier transform of the derivative is given by

$$\widehat{f'}(\xi) = 2\pi i \xi \hat{f}(\xi).$$

More generally, the Fourier transformation of the n-th derivative $f^{(n)}$ is given by

$$\widehat{f^{(n)}}(\xi) = (2\pi i \xi)^n \hat{f}(\xi).$$

By applying the Fourier transform and using these formulas, some ordinary differential equations can be transformed into algebraic equations, which are much easier to solve. These formulas also give rise to the rule of thumb " $f(x)$ is smooth if and only if $\hat{f}(\xi)$ quickly falls down to 0 for $|\xi| \to \infty$." By using the analogous rules for the inverse Fourier transform, one can also say " $f(x)$ quickly falls down to 0 for $|x| \to \infty$ if and only if $\hat{f}(\xi)$ is smooth."

2.5.8   Convolution theorem

Main article: Convolution theorem

The Fourier transform translates between convolution and multiplication of functions. If $f(x)$ and $g(x)$ are integrable functions with Fourier transforms $\hat{f}(\xi)$ and $\hat{g}(\xi)$ respectively, then the Fourier transform of the convolution is given by the product of the Fourier transforms $\hat{f}(\xi)$ and $\hat{g}(\xi)$ (under other conventions for the definition of the Fourier transform a constant factor may appear).

This means that if:

$$h(x) = (f * g)(x) = \int_{-\infty}^{\infty} f(y)g(x-y)\,dy,$$

where $*$ denotes the convolution operation, then:

$$\hat{h}(\xi) = \hat{f}(\xi) \cdot \hat{g}(\xi).$$

In linear time invariant (LTI) system theory, it is common to interpret $g(x)$ as the impulse response of an LTI system with input $f(x)$ and output $h(x)$, since substituting the unit impulse for $f(x)$ yields $h(x) = g(x)$. In this case, $\hat{g}(\xi)$ represents the frequency response of the system.

Conversely, if $f(x)$ can be decomposed as the product of two square integrable functions $p(x)$ and $q(x)$, then the Fourier transform of $f(x)$ is given by the convolution of the respective Fourier transforms $\hat{p}(\xi)$ and $\hat{q}(\xi)$.

2.5.9 Cross-correlation theorem

Main article: Cross-correlation

In an analogous manner, it can be shown that if $h(x)$ is the cross-correlation of $f(x)$ and $g(x)$:

$$h(x) = (f \star g)(x) = \int_{-\infty}^{\infty} \overline{f(y)}\, g(x+y)\, dy$$

then the Fourier transform of $h(x)$ is:

$$\hat{h}(\xi) = \overline{\hat{f}(\xi)} \cdot \hat{g}(\xi).$$

As a special case, the autocorrelation of function $f(x)$ is:

$$h(x) = (f \star f)(x) = \int_{-\infty}^{\infty} \overline{f(y)} f(x+y)\, dy$$

for which

$$\hat{h}(\xi) = \overline{\hat{f}(\xi)}\, \hat{f}(\xi) = |\hat{f}(\xi)|^2.$$

2.5.10 Eigenfunctions

One important choice of an orthonormal basis for $L^2(\mathbf{R})$ is given by the Hermite functions

$$\psi_n(x) = \frac{2^{1/4}}{\sqrt{n!}}\, e^{-\pi x^2} \mathrm{He}_n(2x\sqrt{\pi}),$$

where $\mathrm{He}n(x)$ are the "probabilist's" Hermite polynomials, defined by

$$\mathrm{He}_n(x) = (-1)^n e^{\frac{x^2}{2}} \left(\frac{d}{dx}\right)^n e^{-\frac{x^2}{2}}$$

Under this convention for the Fourier transform, we have that

$$\hat{\psi}_n(\xi) = (-i)^n \psi_n(\xi)$$

In other words, the Hermite functions form a complete orthonormal system of eigenfunctions for the Fourier transform on $L^2(\mathbf{R})$ (Pinsky 2002). However, this choice of eigenfunctions is not unique. There are only four different eigenvalues of the Fourier transform (± 1 and $\pm i$) and any linear combination of eigenfunctions with the same eigenvalue gives another eigenfunction. As a consequence of this, it is possible to decompose $L^2(\mathbf{R})$ as a direct sum of four spaces H_0, H_1, H_2, and H_3 where the Fourier transform acts on Hek simply by multiplication by i^k.

Since the complete set of Hermite functions provides a resolution of the identity, the Fourier transform can be represented by such a sum of terms weighted by the above eigenvalues, and these sums can be explicitly summed. This approach to define the Fourier transform was first done by Norbert Wiener (Duoandikoetxea 2001). Among other properties, Hermite functions decrease exponentially fast in both frequency and time domains, and they are thus used to define a generalization of the Fourier transform, namely the fractional Fourier transform used in time-frequency analysis (Boashash 2003). In physics, this transform was introduced by Edward Condon (Condon 1937).

2.6   Complex domain

The integral for the Fourier transform

$$\hat{f}(\xi) = \int_{-\infty}^{\infty} e^{-2\pi i \xi t} f(t)\, dt$$

can be studied for complex values of its argument ξ. Depending on the properties of f, this might not converge off the real axis at all, or it might converge to a complex analytic function for all values of $\xi = \sigma + i\tau$, or something in between. [4]

The Paley–Wiener theorem says that f is smooth (i.e., n-times differentiable for all positive integers n) and compactly supported if and only if $\hat{f}(\sigma + i\tau)$ is a holomorphic function for which there exists a constant $a > 0$ such that for any integer $n \geq 0$,

$$|\xi^n \hat{f}(\xi)| \leq C e^{a|\tau|}$$

for some constant C. (In this case, f is supported on $[-a, a]$.) This can be expressed by saying that $\hat{f}$ is an entire function which is rapidly decreasing in σ (for fixed τ) and of exponential growth in τ (uniformly in σ). [5]

(If f is not smooth, but only L^2, the statement still holds provided $n = 0$.) [6] The space of such functions of a complex variable is called the Paley—Wiener space. This theorem has been generalised to semi-simple Lie groups. [7]

If f is supported on the half-line $t \geq 0$, then f is said to be "causal" because the impulse response function of a physically realisable filter must have this property, as no effect can precede its cause. Paley and Wiener showed that then $\hat{f}$ extends to a holomorphic function on the complex lower half-plane $\tau < 0$ which tends to zero as τ goes to infinity.[8] The converse is false and it is not known how to characterise the Fourier transform of a causal function. [9]

2.6.1   Laplace transform

The Fourier transform $\hat{f}(\xi)$ is intimately related with the Laplace transform $F(s)$, which is also used for the solution of differential equations and the analysis of filters. Chatfield, indeed, has said that "... the Laplace and the Fourier transforms [of a causal function] are the same, provided that the real part of s is zero." [10]

It may happen that a function f for which the Fourier integral does not converge on the real axis at all, nevertheless has a complex Fourier transform defined in some region of the complex plane.

For example, if $f(t)$ is of exponential growth, i.e.,

$$|f(t)| < C e^{a|t|}$$

for some constants $C, a \geq 0$, then [11]

$$\hat{f}(i\tau) = \int_{-\infty}^{\infty} e^{2\pi\tau t} f(t)\,dt,$$

convergent for all $2\pi\tau < -a$, is the two-sided Laplace transform of f.

The more usual version ("one-sided") of the Laplace transform is

$$F(s) = \int_{0}^{\infty} f(t)e^{-st}\,dt.$$

If f is also causal, then

$$\hat{f}(i\tau) = F(-2\pi\tau).$$

Thus, extending the Fourier transform to the complex domain means it includes the Laplace transform as a special case—the case of causal functions—but with the change of variable $s = 2\pi i\xi$.

2.6.2 Inversion

If $\hat{f}$ has no poles for $a \leq \tau \leq b$, then

$$\int_{-\infty}^{\infty} \hat{f}(\sigma + ia)e^{2\pi i\xi t}\,d\sigma = \int_{-\infty}^{\infty} \hat{f}(\sigma + ib)e^{2\pi i\xi t}\,d\sigma$$

by Cauchy's integral theorem. Therefore, the Fourier inversion formula can use integration along different lines, parallel to the real axis. [12]

Theorem: If $f(t) = 0$ for $t < 0$, and $|f(t)| < Ce^{a|t|}$ for some constants $C, a > 0$, then

$$f(t) = \int_{-\infty}^{\infty} \hat{f}(\sigma + i\tau)e^{2\pi i\xi t}\,d\sigma,$$

for any $\tau < -\frac{a}{2\pi}$.

This theorem implies the Mellin inversion formula for the Laplace transformation, [13]

$$f(t) = \frac{1}{2\pi i}\int_{b-i\infty}^{b+i\infty} F(s)e^{st}\,ds$$

for any $b > a$, where $F(s)$ is the Laplace transform of $f(t)$.

The hypotheses can be weakened, as in the results of Carleman and Hunt, to $f(t)e^{-at}$ being L^1, provided that t is in the interior of a closed interval on which f is continuous and of bounded variation, and provided that the integrals are taken in the sense of Cauchy principal values. [14]

L^2 versions of these inversion formulas are also available.[15]

2.7 Fourier transform on Euclidean space

The Fourier transform can be defined in any arbitrary number of dimensions n. As with the one-dimensional case, there are many conventions. For an integrable function $f(\mathbf{x})$, this article takes the definition:

$$\hat{f}(\boldsymbol{\xi}) = \mathcal{F}(f)(\boldsymbol{\xi}) = \int_{\mathbb{R}^n} f(\mathbf{x}) e^{-2\pi i \mathbf{x} \cdot \boldsymbol{\xi}} \, d\mathbf{x}$$

where $\mathbf{x}$ and $\boldsymbol{\xi}$ are n-dimensional vectors, and $\mathbf{x} \cdot \boldsymbol{\xi}$ is the dot product of the vectors. The dot product is sometimes written as $\langle \mathbf{x}, \boldsymbol{\xi} \rangle$.

All of the basic properties listed above hold for the n-dimensional Fourier transform, as do Plancherel's and Parseval's theorem. When the function is integrable, the Fourier transform is still uniformly continuous and the Riemann–Lebesgue lemma holds. (Stein & Weiss 1971)

2.7.1   Uncertainty principle

For more details on this topic, see Gabor limit.

Generally speaking, the more concentrated $f(x)$ is, the more spread out its Fourier transform $\hat{f}(\xi)$ must be. In particular, the scaling property of the Fourier transform may be seen as saying: if we "squeeze" a function in x, its Fourier transform "stretches out" in ξ. It is not possible to arbitrarily concentrate both a function and its Fourier transform.

The trade-off between the compaction of a function and its Fourier transform can be formalized in the form of an **uncertainty principle** by viewing a function and its Fourier transform as conjugate variables with respect to the symplectic form on the time–frequency domain: from the point of view of the linear canonical transformation, the Fourier transform is rotation by 90° in the time–frequency domain, and preserves the symplectic form.

Suppose $f(x)$ is an integrable and square-integrable function. Without loss of generality, assume that $f(x)$ is normalized:

$$\int_{-\infty}^{\infty} |f(x)|^2 \, dx = 1.$$

It follows from the Plancherel theorem that $\hat{f}(\xi)$ is also normalized.

The spread around $x = 0$ may be measured by the *dispersion about zero* (Pinsky 2002, p. 131) defined by

$$D_0(f) = \int_{-\infty}^{\infty} x^2 |f(x)|^2 \, dx.$$

In probability terms, this is the second moment of $|f(x)|^2$ about zero.

The Uncertainty principle states that, if $f(x)$ is absolutely continuous and the functions $x \cdot f(x)$ and $f'(x)$ are square integrable, then

$$D_0(f) D_0(\hat{f}) \geq \tfrac{1}{16\pi^2} \text{ (Pinsky 2002)}.$$

The equality is attained only in the case $f(x) = C_1 \, e^{-\pi x^2/\sigma^2}$ (hence $\hat{f}(\xi) = \sigma C_1 \, e^{-\pi \sigma^2 \xi^2}$) where $\sigma > 0$ is arbitrary and $C_1 = \sqrt[4]{2}/\sqrt{\sigma}$ so that f is L^2–normalized (Pinsky 2002). In other words, where f is a (normalized) Gaussian function with variance σ^2, centered at zero, and its Fourier transform is a Gaussian function with variance σ^{-2}.

In fact, this inequality implies that:

$$\left(\int_{-\infty}^{\infty} (x - x_0)^2 |f(x)|^2 \, dx \right) \left(\int_{-\infty}^{\infty} (\xi - \xi_0)^2 |\hat{f}(\xi)|^2 \, d\xi \right) \geq \frac{1}{16\pi^2}$$

for any $x_0, \xi_0 \in \mathbf{R}$ (Stein & Shakarchi 2003, p. 158).

In quantum mechanics, the momentum and position wave functions are Fourier transform pairs, to within a factor of Planck's constant. With this constant properly taken into account, the inequality above becomes the statement of the Heisenberg uncertainty principle (Stein & Shakarchi 2003, p. 158).

A stronger uncertainty principle is the Hirschman uncertainty principle, which is expressed as:

$$H(|f|^2) + H(|\hat{f}|^2) \geq \log(e/2)$$

where $H(p)$ is the differential entropy of the probability density function $p(x)$:

$$H(p) = -\int_{-\infty}^{\infty} p(x) \log(p(x))\, dx$$

where the logarithms may be in any base that is consistent. The equality is attained for a Gaussian, as in the previous case.

2.7.2 Sine and cosine transforms

Main article: Sine and cosine transforms

Fourier's original formulation of the transform did not use complex numbers, but rather sines and cosines. Statisticians and others still use this form. An absolutely integrable function f for which Fourier inversion holds good can be expanded in terms of genuine frequencies (avoiding negative frequencies, which are sometimes considered hard to interpret physically[16]) λ by

$$f(t) = \int_0^\infty [a(\lambda)\cos 2\pi\lambda t + b(\lambda)\sin 2\pi\lambda t]\ d\lambda.$$

This is called an expansion as a trigonometric integral, or a Fourier integral expansion. The coefficient functions a and b can be found by using variants of the Fourier cosine transform and the Fourier sine transform (the normalisations are, again, not standardised):

$$a(\lambda) = 2\int_{-\infty}^{\infty} f(t)\cos 2\pi\lambda t\, dt$$

and

$$b(\lambda) = 2\int_{-\infty}^{\infty} f(t)\sin 2\pi\lambda t\, dt.$$

Older literature refers to the two transform functions, the Fourier cosine transform, a, and the Fourier sine transform, b.

The function f can be recovered from the sine and cosine transform using

$$f(t) = 2\int_0^\infty \int_{-\infty}^{\infty} f(\tau)\cos 2\pi\lambda(\tau - t)\, d\tau\, d\lambda.$$

together with trigonometric identities. This is referred to as Fourier's integral formula.[17]

2.7.3 Spherical harmonics

Let the set of homogeneous harmonic polynomials of degree k on $\mathbf{R}^n$ be denoted by $\mathbf{A}k$. The set $\mathbf{A}k$ consists of the solid spherical harmonics of degree k. The solid spherical harmonics play a similar role in higher dimensions to the Hermite polynomials in dimension one. Specifically, if $f(x) = e^{-\pi|x|^2}P(x)$ for some $P(x)$ in $\mathbf{A}k$, then $\hat{f}(\xi) = i^{-k}f(\xi)$. Let the set $\mathbf{H}k$ be the closure in $L^2(\mathbf{R}^n)$ of linear combinations of functions of the form $f(|x|)P(x)$ where $P(x)$ is in $\mathbf{A}k$. The space $L^2(\mathbf{R}^n)$ is then a direct sum of the spaces $\mathbf{H}k$ and the Fourier transform maps each space $\mathbf{H}k$ to itself and is possible to characterize the action of the Fourier transform on each space $\mathbf{H}k$ (Stein & Weiss 1971). Let $f(x) = f_0(|x|)P(x)$ (with $P(x)$ in $\mathbf{A}k$), then $\hat{f}(\xi) = F_0(|\xi|)P(\xi)$ where

$$F_0(r) = 2\pi i^{-k}r^{-(n+2k-2)/2} \int_0^\infty f_0(s)J_{(n+2k-2)/2}(2\pi rs)s^{(n+2k)/2}\, ds.$$

Here $J_{(n+2k-2)/2}$ denotes the Bessel function of the first kind with order $(n+2k-2)/2$. When $k = 0$ this gives a useful formula for the Fourier transform of a radial function (Grafakos 2004). Note that this is essentially the Hankel transform. Moreover, there is a simple recursion relating the cases $n + 2$ and n (Grafakos & Teschl 2013) allowing to compute, e.g., the three-dimensional Fourier transform of a radial function from the one-dimensional one.

2.7.4 Restriction problems

In higher dimensions it becomes interesting to study *restriction problems* for the Fourier transform. The Fourier transform of an integrable function is continuous and the restriction of this function to any set is defined. But for a square-integrable function the Fourier transform could be a general *class* of square integrable functions. As such, the restriction of the Fourier transform of an $L^2(\mathbf{R}^n)$ function cannot be defined on sets of measure 0. It is still an active area of study to understand restriction problems in L^p for $1 < p < 2$. Surprisingly, it is possible in some cases to define the restriction of a Fourier transform to a set S, provided S has non-zero curvature. The case when S is the unit sphere in $\mathbf{R}^n$ is of particular interest. In this case the Tomas–Stein restriction theorem states that the restriction of the Fourier transform to the unit sphere in $\mathbf{R}^n$ is a bounded operator on L^p provided $1 \le p \le (2n + 2) / (n + 3)$.

One notable difference between the Fourier transform in 1 dimension versus higher dimensions concerns the partial sum operator. Consider an increasing collection of measurable sets ER indexed by $R \in (0,\infty)$: such as balls of radius R centered at the origin, or cubes of side $2R$. For a given integrable function f, consider the function fR defined by:

$$f_R(x) = \int_{E_R} \hat{f}(\xi) e^{2\pi i x \cdot \xi}\, d\xi, \quad x \in \mathbf{R}^n.$$

Suppose in addition that $f \in L^p(\mathbf{R}^n)$. For $n = 1$ and $1 < p < \infty$, if one takes $ER = (-R, R)$, then fR converges to f in L^p as R tends to infinity, by the boundedness of the Hilbert transform. Naively one may hope the same holds true for $n > 1$. In the case that ER is taken to be a cube with side length R, then convergence still holds. Another natural candidate is the Euclidean ball $ER = \{\xi : |\xi| < R\}$. In order for this partial sum operator to converge, it is necessary that the multiplier for the unit ball be bounded in $L^p(\mathbf{R}^n)$. For $n \ge 2$ it is a celebrated theorem of Charles Fefferman that the multiplier for the unit ball is never bounded unless $p = 2$ (Duoandikoetxea 2001). In fact, when $p \ne 2$, this shows that not only may fR fail to converge to f in L^p, but for some functions $f \in L^p(\mathbf{R}^n)$, fR is not even an element of L^p.

2.8   Fourier transform on function spaces

2.8.1   On L^p spaces

On L^1

The definition of the Fourier transform by the integral formula

$$\hat{f}(\xi) = \int_{\mathbf{R}^n} f(x) e^{-2\pi i \xi \cdot x}\, dx$$

is valid for Lebesgue integrable functions f; that is, $f \in L^1(\mathbf{R}^n)$.

The Fourier transform $\mathcal{F} : L^1(\mathbf{R}^n) \to L^\infty(\mathbf{R}^n)$ is a bounded operator. This follows from the observation that

$$|\hat{f}(\xi)| \le \int_{\mathbf{R}^n} |f(x)|\, dx,$$

which shows that its operator norm is bounded by 1. Indeed, it equals 1, which can be seen, for example, from the transform of the rect function. The image of L^1 is a subset of the space $C_0(\mathbf{R}^n)$ of continuous functions that tend to zero at infinity (the Riemann–Lebesgue lemma), although it is not the entire space. Indeed, there is no simple characterization of the image.

On L^2

Since compactly supported smooth functions are integrable and dense in $L^2(\mathbf{R}^n)$, the Plancherel theorem allows us to extend the definition of the Fourier transform to general functions in $L^2(\mathbf{R}^n)$ by continuity arguments. The Fourier transform in $L^2(\mathbf{R}^n)$ is no longer given by an ordinary Lebesgue integral, although it can be computed by an improper integral, here meaning that for an L^2 function f,

$$\hat{f}(\xi) = \lim_{R \to \infty} \int_{|x| \leq R} f(x) e^{-2\pi i x \cdot \xi} \, dx$$

where the limit is taken in the L^2 sense. Many of the properties of the Fourier transform in L^1 carry over to L^2, by a suitable limiting argument.

Furthermore, $\mathcal{F} : L^2(\mathbf{R}^n) \to L^2(\mathbf{R}^n)$ is a unitary operator (Stein & Weiss 1971, Thm. 2.3). For an operator to be unitary it is sufficient to show that it is bijective and preserves the inner product, so in this case these follow from the Fourier inversion theorem combined with the fact that for any $f, g \in L^2(\mathbf{R}^n)$ we have

$$\int_{\mathbf{R}^n} f(x) \mathcal{F} g(x) \, dx = \int_{\mathbf{R}^n} \mathcal{F} f(x) g(x) \, dx.$$

In particular, the image of $L^2(\mathbf{R}^n)$ is itself under the Fourier transform.

On other L^p

The definition of the Fourier transform can be extended to functions in $L^p(\mathbf{R}^n)$ for $1 \leq p \leq 2$ by decomposing such functions into a fat tail part in L^2 plus a fat body part in L^1. In each of these spaces, the Fourier transform of a function in $L^p(\mathbf{R}^n)$ is in $L^q(\mathbf{R}^n)$, where $q = p/(p-1)$ is the Hölder conjugate of p. by the Hausdorff–Young inequality. However, except for $p = 2$, the image is not easily characterized. Further extensions become more technical. The Fourier transform of functions in L^p for the range $2 < p < \infty$ requires the study of distributions (Katznelson 1976). In fact, it can be shown that there are functions in L^p with $p > 2$ so that the Fourier transform is not defined as a function (Stein & Weiss 1971).

2.8.2 Tempered distributions

Main article: Distribution (mathematics) § Tempered distributions and Fourier transform

One might consider enlarging the domain of the Fourier transform from $L^1 + L^2$ by considering generalized functions, or distributions. A distribution on $\mathbf{R}^n$ is a continuous linear functional on the space $C_c(\mathbf{R}^n)$ of compactly supported smooth functions, equipped with a suitable topology. The strategy is then to consider the action of the Fourier transform on $C_c(\mathbf{R}^n)$ and pass to distributions by duality. The obstruction to do this is that the Fourier transform does not map $C_c(\mathbf{R}^n)$ to $C_c(\mathbf{R}^n)$. In fact the Fourier transform of an element in $C_c(\mathbf{R}^n)$ can not vanish on an open set; see the above discussion on the uncertainty principle. The right space here is the slightly larger space of Schwartz functions. The Fourier transform is an automorphism on the Schwartz space, as a topological vector space, and thus induces an automorphism on its dual, the space of tempered distributions (Stein & Weiss 1971). The tempered distributions include all the integrable functions mentioned above, as well as well-behaved functions of polynomial growth and distributions of compact support.

For the definition of the Fourier transform of a tempered distribution, let f and g be integrable functions, and let $\hat{f}$ and $\hat{g}$ be their Fourier transforms respectively. Then the Fourier transform obeys the following multiplication formula (Stein & Weiss 1971),

$$\int_{\mathbf{R}^n} \hat{f}(x) g(x) \, dx = \int_{\mathbf{R}^n} f(x) \hat{g}(x) \, dx.$$

Every integrable function f defines (induces) a distribution Tf by the relation

$T_f(\varphi) = \int_{\mathbf{R}^n} f(x)\varphi(x)\,dx$ for all Schwartz functions φ.

So it makes sense to define Fourier transform $\hat{T}_f$ of Tf by

$$\hat{T}_f(\varphi) = T_f(\hat{\varphi})$$

for all Schwartz functions φ. Extending this to all tempered distributions T gives the general definition of the Fourier transform.

Distributions can be differentiated and the above-mentioned compatibility of the Fourier transform with differentiation and convolution remains true for tempered distributions.

2.9 Generalizations

2.9.1 Fourier–Stieltjes transform

The Fourier transform of a finite Borel measure μ on $\mathbf{R}^n$ is given by (Pinsky 2002, p. 256):

$$\hat{\mu}(\xi) = \int_{\mathbf{R}^n} e^{-2\pi i x \cdot \xi}\,d\mu.$$

This transform continues to enjoy many of the properties of the Fourier transform of integrable functions. One notable difference is that the Riemann–Lebesgue lemma fails for measures (Katznelson 1976). In the case that $d\mu = f(x)\,dx$, then the formula above reduces to the usual definition for the Fourier transform of f. In the case that μ is the probability distribution associated to a random variable X, the Fourier–Stieltjes transform is closely related to the characteristic function, but the typical conventions in probability theory take $e^{ix \cdot \xi}$ instead of $e^{-2\pi i x \cdot \xi}$ (Pinsky 2002). In the case when the distribution has a probability density function this definition reduces to the Fourier transform applied to the probability density function, again with a different choice of constants.

The Fourier transform may be used to give a characterization of measures. Bochner's theorem characterizes which functions may arise as the Fourier–Stieltjes transform of a positive measure on the circle (Katznelson 1976).

Furthermore, the Dirac delta function is not a function but it is a finite Borel measure. Its Fourier transform is a constant function (whose specific value depends upon the form of the Fourier transform used).

2.9.2 Locally compact abelian groups

Main article: Pontryagin duality

The Fourier transform may be generalized to any locally compact abelian group. A locally compact abelian group is an abelian group that is at the same time a locally compact Hausdorff topological space so that the group operation is continuous. If G is a locally compact abelian group, it has a translation invariant measure μ, called Haar measure. For a locally compact abelian group G, the set of irreducible, i.e. one-dimensional, unitary representations are called its characters. With its natural group structure and the topology of pointwise convergence, the set of characters $\hat{G}$ is itself a locally compact abelian group, called the *Pontryagin dual* of G. For a function f in $L^1(G)$, its Fourier transform is defined by (Katznelson 1976):

$$\hat{f}(\xi) = \int_G \xi(x) f(x)\,d\mu \qquad \text{any for} \xi \in \hat{G}.$$

The Riemann–Lebesgue lemma holds in this case; $\hat{f}(\xi)$ is a function vanishing at infinity on $\hat{G}$.

2.9.3   Gelfand transform

Main article: Gelfand representation

The Fourier transform is also a special case of Gelfand transform. In this particular context, it is closely related to the Pontryagin duality map defined above.

Given an abelian locally compact Hausdorff topological group G, as before we consider space $L^1(G)$, defined using a Haar measure. With convolution as multiplication, $L^1(G)$ is an abelian Banach algebra. It also has an involution * given by

$$f^*(g) = \overline{f(g^{-1})}.$$

Taking the completion with respect to the largest possibly C*-norm gives its enveloping C*-algebra, called the group C*-algebra C*(G) of G. (Any C*-norm on $L^1(G)$ is bounded by the L^1 norm, therefore their supremum exists.)

Given any abelian C*-algebra A, the Gelfand transform gives an isomorphism between A and $C_0(A^\wedge)$, where $A^\wedge$ is the multiplicative linear functionals, i.e. one-dimensional representations, on A with the weak-* topology. The map is simply given by

$$a \mapsto (\varphi \mapsto \varphi(a))$$

It turns out that the multiplicative linear functionals of $C^*(G)$, after suitable identification, are exactly the characters of G, and the Gelfand transform, when restricted to the dense subset $L^1(G)$ is the Fourier-Pontryagin transform.

2.9.4   Compact non-abelian groups

The Fourier transform can also be defined for functions on a non-abelian group, provided that the group is compact. Removing the assumption that the underlying group is abelian, irreducible unitary representations need not always be one-dimensional. This means the Fourier transform on a non-abelian group takes values as Hilbert space operators (Hewitt & Ross 1970, Chapter 8). The Fourier transform on compact groups is a major tool in representation theory (Knapp 2001) and non-commutative harmonic analysis.

Let G be a compact Hausdorff topological group. Let Σ denote the collection of all isomorphism classes of finite-dimensional irreducible unitary representations, along with a definite choice of representation $U^{(\sigma)}$ on the Hilbert space $H\sigma$ of finite dimension $d\sigma$ for each $\sigma \in \Sigma$. If μ is a finite Borel measure on G, then the Fourier–Stieltjes transform of μ is the operator on $H\sigma$ defined by

$$\langle \hat{\mu}\xi, \eta \rangle_{H_\sigma} = \int_G \langle \overline{U}_g^{(\sigma)} \xi, \eta \rangle \, d\mu(g)$$

where $\overline{U}^{(\sigma)}$ is the complex-conjugate representation of $U^{(\sigma)}$ acting on $H\sigma$. If μ is absolutely continuous with respect to the left-invariant probability measure λ on G, represented as

$$d\mu = f \, d\lambda$$

for some $f \in L^1(\lambda)$, one identifies the Fourier transform of f with the Fourier–Stieltjes transform of μ.

The mapping $\mu \mapsto \hat{\mu}$ defines an isomorphism between the Banach space $M(G)$ of finite Borel measures (see rca space) and a closed subspace of the Banach space $\mathbf{C}\infty(\Sigma)$ consisting of all sequences $E = (E\sigma)$ indexed by Σ of (bounded) linear operators $E\sigma: H\sigma \to H\sigma$ for which the norm

$$\|E\| = \sup_{\sigma \in \Sigma} \|E_\sigma\|$$

is finite. The "convolution theorem" asserts that, furthermore, this isomorphism of Banach spaces is in fact an isometric isomorphism of C* algebras into a subspace of $\mathbf{C}\infty(\Sigma)$. Multiplication on $M(G)$ is given by convolution of measures and the involution * defined by

$$f^*(g) = \overline{f(g^{-1})},$$

and $\mathbf{C}\infty(\Sigma)$ has a natural C*-algebra structure as Hilbert space operators.

The Peter–Weyl theorem holds, and a version of the Fourier inversion formula (Plancherel's theorem) follows: if $f \in L^2(G)$, then

$$f(g) = \sum_{\sigma \in \Sigma} d_\sigma \operatorname{tr}(\hat{f}(\sigma) U_g^{(\sigma)})$$

where the summation is understood as convergent in the L^2 sense.

The generalization of the Fourier transform to the noncommutative situation has also in part contributed to the development of noncommutative geometry. In this context, a categorical generalization of the Fourier transform to noncommutative groups is Tannaka–Krein duality, which replaces the group of characters with the category of representations. However, this loses the connection with harmonic functions.

2.10   Alternatives

In signal processing terms, a function (of time) is a representation of a signal with perfect *time resolution*, but no frequency information, while the Fourier transform has perfect *frequency resolution*, but no time information: the magnitude of the Fourier transform at a point is how much frequency content there is, but location is only given by phase (argument of the Fourier transform at a point), and standing waves are not localized in time – a sine wave continues out to infinity, without decaying. This limits the usefulness of the Fourier transform for analyzing signals that are localized in time, notably transients, or any signal of finite extent.

As alternatives to the Fourier transform, in time-frequency analysis, one uses time-frequency transforms or time-frequency distributions to represent signals in a form that has some time information and some frequency information – by the uncertainty principle, there is a trade-off between these. These can be generalizations of the Fourier transform, such as the short-time Fourier transform or fractional Fourier transform, or other functions to represent signals, as in wavelet transforms and chirplet transforms, with the wavelet analog of the (continuous) Fourier transform being the continuous wavelet transform. (Boashash 2003).

2.11   Applications

See also: Spectral density § Applications

2.11.1   Analysis of differential equations

Perhaps the most important use of the Fourier transformation is to solve partial differential equations. Many of the equations of the mathematical physics of the nineteenth century can be treated this way. Fourier studied the heat equation, which in one dimension and in dimensionless units is

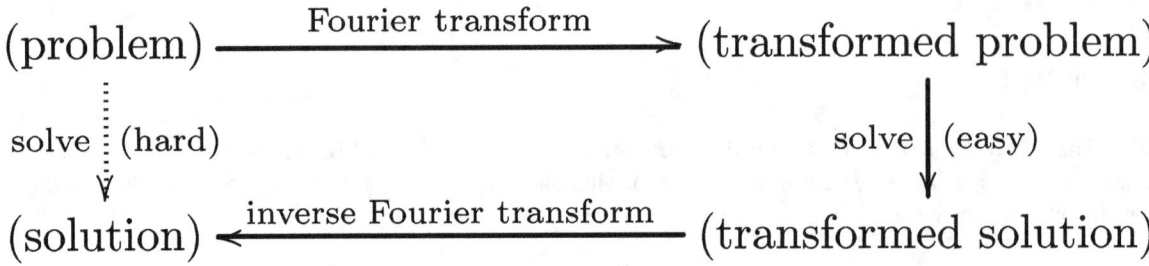

Some problems, such as certain differential equations, become easier to solve when the Fourier transform is applied. In that case the solution to the original problem is recovered using the inverse Fourier transform.

$\frac{\partial^2 y(x,t)}{\partial^2 x} = \frac{\partial y(x,t)}{\partial t}.$

The example we will give, a slightly more difficult one, is the wave equation in one dimension,

$\frac{\partial^2 y(x,t)}{\partial^2 x} = \frac{\partial^2 y(x,t)}{\partial^2 t}.$

As usual, the problem is not to find a solution: there are infinitely many. The problem is that of the so-called "boundary problem": find a solution which satisfies the "boundary conditions"

$$y(x,0) = f(x), \frac{\partial y(x,t)}{\partial t} = g(x).$$

Here, f and g are given functions. For the heat equation, only one boundary condition can be required (usually the first one). But for the wave equation, there are still infinitely many solutions y which satisfy the first boundary condition. But when one imposes both conditions, there is only one possible solution.

It is easier to find the Fourier transform $\hat{y}$ of the solution than to find the solution directly. This is because the Fourier transformation takes differentiation into multiplication by the variable, and so a partial differential equation applied to the original function is transformed into multiplication by polynomial functions of the dual variables applied to the transformed function. After $\hat{y}$ is determined, we can apply the inverse Fourier transformation to find y.

Fourier's method is as follows. First, note that any function of the forms

$\cos 2\pi\xi(x \pm t)$ or $\sin 2\pi\xi(x \pm t)$

satisfies the wave equation. These are called "the elementary solutions."

Second, note that therefore any integral

$$y(x,t) = \int_0^\infty a_+(\xi)\cos 2\pi\xi(x+t) + a_-(\xi)\cos 2\pi\xi(x-t) + b_+(\xi)\sin 2\pi\xi(x+t) + b_-(\xi)\sin 2\pi\xi(x-t)\, d\xi$$

(for arbitrary a_+, a_-, b_+, and b_-) satisfies the wave equation. (This integral is just a kind of continuous linear combination, and the equation is linear.)

Now this resembles the formula for the Fourier synthesis of a function. In fact, this is the real inverse Fourier transform of $a_\pm$ and $b_\pm$ in the variable x.

The third step is to examine how to find the specific unknown coefficient functions $a_\pm$ and $b_\pm$ that will lead to y's satisfying the boundary conditions. We are interested in the values of these solutions at $t = 0$. So we will set $t = 0$. Assuming that the conditions needed for Fourier inversion are satisfied, we can then find the Fourier sine and cosine transforms (in the variable x) of both sides and obtain

$$2\int_{-\infty}^\infty y(x,0)\cos 2\pi\xi x\, dx = a_+ + a_-$$

and

$$2 \int_{-\infty}^{\infty} y(x,0) \sin 2\pi\xi x \, dx = b_+ + b_-.$$

Similarly, taking the derivative of y with respect to t and then applying the Fourier sine and cosine transformations yields

$$2 \int_{-\infty}^{\infty} \frac{\partial y(u,0)}{\partial t} \sin(2\pi\xi x) \, dx = (2\pi\xi)(-a_+ + a_-)$$

and

$$2 \int_{-\infty}^{\infty} \frac{\partial y(u,0)}{\partial t} \cos(2\pi\xi x) \, dx = (2\pi\xi)(b_+ - b_-).$$

These are four linear equations for the four unknowns $a_\pm$ and $b_\pm$, in terms of the Fourier sine and cosine transforms of the boundary conditions, which are easily solved by elementary algebra, provided that these transforms can be found.

In summary, we chose a set of elementary solutions, parametrised by ξ , of which the general solution would be a (continuous) linear combination in the form of an integral over the parameter ξ . But this integral was in the form of a Fourier integral. The next step was to express the boundary conditions in terms of these integrals, and set them equal to the given functions f and g . But these expressions also took the form of a Fourier integral because of the properties of the Fourier transform of a derivative. The last step was to exploit Fourier inversion by applying the Fourier transformation to both sides, thus obtaining expressions for the coefficient functions $a_\pm$ and $b_\pm$ in terms of the given boundary conditions f and g .

George Mackey has called this procedure "Fourier's algorithm." "Harmonic Analysis as the Exploitation of Symmetry—A Historical Survey." Rice University Studies Volume 64 (1978), Numbers 2 and 3, pp. 73–228. Reprinted in Bulletin of the American Mathematical Society (New Series) Volume 3 (1980), Number 1 (July—November), pp. 543–698.

In fact, one can perform further manipulations on the answer obtained, but those belong rather to the subject of partial differential equations than to the subject of the Fourier transformation itself.

From a higher point of view, Fourier's procedure can be reformulated more conceptually. Since there are two variables, we will use the Fourier transformation in both x and t rather than operate as Fourier did, who only transformed in the spatial variables. Note that $\hat{y}$ must be considered in the sense of a distribution since $y(x,t)$ is not going to be L^1 : as a wave, it will persist through time and thus is not a transient phenomenon. But it will be bounded and so its Fourier transform can be defined as a distribution. The operational properties of the Fourier transformation that are relevant to this equation are that it takes differentiation in x to multiplication by $2\pi i\xi$ and differentiation with respect to t to multiplication by $2\pi i f$ where f is the frequency. Then the wave equation becomes an algebraic equation in $\hat{y}$:

$$\xi^2 \hat{y}(\xi, f) = f^2 \hat{y}(\xi, f).$$

This is equivalent to requiring $\hat{y}(\xi, f) = 0$ unless $\xi = \pm f$. Right away, this explains why the choice of elementary solutions we made earlier worked so well: obviously $\hat{f} = \delta(\xi \pm f)$ will be solutions. Applying Fourier inversion to these delta functions, we obtain the elementary solutions we picked earlier. But from the higher point of view, one does not pick elementary solutions, but rather considers the space of all distributions which are supported on the (degenerate) conic $\xi^2 - f^2 = 0$.

We may as well consider the distributions supported on the conic that are given by distributions of one variable on the line $\xi = f$ plus distributions on the line $\xi = -f$ as follows: if ϕ is any test function,

$$\iint \hat{y}\phi(\xi, f) \, d\xi \, df = \int s_+\phi(\xi, \xi) \, d\xi + \int s_-\phi(\xi, -\xi) \, d\xi,$$

where s_+ , and s_- , are distributions of one variable.

Then Fourier inversion gives, for the boundary conditions, something very similar to what we had more concretely above (put $\phi(\xi, f) = e^{2\pi i(x\xi + tf)}$, which is clearly of polynomial growth):

$$y(x, 0) = \int \{s_+(\xi) + s_-(\xi)\} e^{2\pi i\xi x + 0} \, d\xi$$

and

$$\frac{\partial y(x, 0)}{\partial t} = \int \{s_+(\xi) - s_-(\xi)\} 2\pi i\xi e^{2\pi i\xi x + 0} \, d\xi.$$

Now, as before, applying the one-variable Fourier transformation in the variable x to these functions of x yields two equations in the two unknown distributions $s_\pm$ (which can be taken to be ordinary functions if the boundary conditions are L^1 or L^2).

From a calculational point of view, the drawback of course is that one must first calculate the Fourier transforms of the boundary conditions, then assemble the solution from these, and then calculate an inverse Fourier transform. Closed form formulas are rare, except when there is some geometric symmetry that can be exploited, and the numerical calculations are difficult because of the oscillatory nature of the integrals, which makes convergence slow and hard to estimate. For practical calculations, other methods are often used.

The twentieth century has seen the extension of these methods to all linear partial differential equations with polynomial coefficients, and by extending the notion of Fourier transformation to include Fourier integral operators, some non-linear equations as well.

2.11.2   Fourier transform spectroscopy

Main article: Fourier transform spectroscopy

The Fourier transform is also used in nuclear magnetic resonance (NMR) and in other kinds of spectroscopy, e.g. infrared (FTIR). In NMR an exponentially shaped free induction decay (FID) signal is acquired in the time domain and Fourier-transformed to a Lorentzian line-shape in the frequency domain. The Fourier transform is also used in magnetic resonance imaging (MRI) and mass spectrometry.

2.11.3   Quantum mechanics

The Fourier transform is useful in Quantum Mechanics in two different ways. To begin with, the basic conceptual structure of Quantum Mechanics postulates the existence of pairs of complementary variables, connected by the Heisenberg uncertainty principle. For example, in one dimension, the spatial variable q of, say, a particle, can only be measured by the quantum mechanical "position operator" at the cost of losing information about the momentum p of the particle. Therefore, the physical state of the particle can either be described by a function, called "the wave function", of q or by a function of p but not by a function of both variables. The variable p is called the conjugate variable to q . In Classical Mechanics, the physical state of a particle (existing in one dimension, for simplicity of exposition) would be given by assigning definite values to both p and q simultaneously. Thus, the set of all possible physical states is the two-dimensional real vector space with a p -axis and a q -axis.

In contrast, quantum mechanics chooses a polarisation of this space in the sense that it picks a subspace of one-half the dimension, for example, the q -axis alone, but instead of considering only points, takes the set of all complex-valued "wave functions" on this axis. Nevertheless, choosing the p -axis is an equally valid polarisation, yielding a different representation of the set of possible physical states of the particle which is related to the first representation by the Fourier transformation

$\phi(p) = \int \psi(q) e^{\frac{2\pi i p q}{h}} \, dq.$

Physically realisable states are L^2, and so by the Plancherel theorem, their Fourier transforms are also L^2. (Note that since q is in units of distance and p is in units of momentum, the presence of Planck's constant in the exponent makes the exponent dimensionless, as it should be.)

Therefore, the Fourier transform can be used to pass from one way of representing the state of the particle, by a wave function of position, to another way of representing the state of the particle: by a wave function of momentum. Infinitely many different polarisations are possible, and all are equally valid. Being able to transform states from one representation to another is sometimes convenient.

The other use of the Fourier transform in both Quantum Mechanics and Quantum Field Theory is to solve the applicable wave equation. In non-relativistic Quantum Mechanics, Schroedinger's equation for a time-varying wave function in one-dimension, not subject to external forces, is

$\frac{\partial^2}{\partial x^2} \psi(x,t) = i \frac{h}{2\pi} \frac{\partial}{\partial t} \psi(x,t).$

This is the same as the heat equation except for the presence of the imaginary unit i. Fourier methods can be used to solve this equation.

In the presence of a potential, given by the potential energy function $V(x)$, the equation becomes

$\frac{\partial^2}{\partial x^2} \psi(x,t) + V(x)\psi(x,t) = i \frac{h}{2\pi} \frac{\partial}{\partial t} \psi(x,t).$

The "elementary solutions", as we referred to them above, are the so-called "stationary states" of the particle, and Fourier's algorithm, as described above, can still be used to solve the boundary value problem of the future evolution of ψ given its values for $t = 0$. Neither of these approaches is of much practical use in Quantum Mechanics. Boundary value problems and the time-evolution of the wave function is not of much practical interest: it is the stationary states that are most important.

In relativistic Quantum Mechanics, Schroedinger's equation becomes a wave equation as was usual in classical physics, except that complex-valued waves are considered. A simple example, in the absence of interactions with other particles or fields, is the free one-dimensional Klein—Gordon—Schroedinger—Fock equation, this time in dimensionless units,

$(\frac{\partial^2}{\partial x^2} + 1)\psi(x,t) = \frac{\partial^2}{\partial t^2} \psi(x,t).$

This is, from the mathematical point of view, the same as the wave equation of classical physics solved above (but with a complex-valued wave, which makes no difference in the methods). This is of great use in Quantum Field Theory: each separate Fourier component of a wave can be treated as a separate harmonic oscillator and then quantised, a procedure known as "second quantisation". Fourier methods have been adapted to also deal with non-trivial interactions.

2.11.4   Signal processing

The Fourier transform is used for the spectral analysis of time-series. The subject of statistical signal processing does not, however, usually apply the Fourier transformation to the signal itself. Even if a real signal is indeed transient, it has been found in practice advisable to model a signal by a function (or, alternatively, a stochastic process) which is stationary in the sense that its characteristic properties are constant over all time. The Fourier transform of such a function does not exist in the usual sense, and it has been found more useful for the analysis of signals to instead take the Fourier transform of its auto-correlation function.

The auto-correlation function R of a function f is defined by

$\mathbb{R}_f(\tau) = \lim_{T \to \infty} \frac{1}{2T} \int_{-T}^{T} f(t)f(t+\tau) \, dt.$

This function is a function of the time-lag τ elapsing between the values of f to be correlated.

For most functions f that occur in practice, R is a bounded even function of the time-lag τ and for typical noisy signals it turns out to be uniformly continuous with a maximum at $\tau = $ zero.

The auto-correlation function, more properly called the auto-covariance function unless it is normalised in some appropriate fashion, measures the strength of the correlation between the values of f separated by a time-lag. This is a way of searching for the correlation of f with its own past. It is useful even for other statistical tasks besides the analysis of signals. For example, if $f(t)$ represents the temperature at time t, one expects a strong correlation with the temperature at a time-lag of 24 hours.

It possesses a Fourier transform,

$$P_f(\xi) = \int_{-\infty}^{\infty} R_f(\tau) e^{-2\pi i \xi \tau} \, d\tau.$$

This Fourier transform is called the power spectral density function of f. (Unless all periodic components are first filtered out from f, this integral will diverge, but it is easy to filter out such periodicities.)

The power spectrum, as indicated by this density function P, measures the amount of variance contributed to the data by the frequency ξ. In electrical signals, the variance is proportional to the average power (energy per unit time), and so the power spectrum describes how much the different frequencies contribute to the average power of the signal. This process is called the spectral analysis of time-series and is analogous to the usual analysis of variance of data that is not a time-series (ANOVA).

Knowledge of which frequencies are "important" in this sense is crucial for the proper design of filters and for the proper evaluation of measuring apparatuses. It can also be useful for the scientific analysis of the phenomena responsible for producing the data.

The power spectrum of a signal can also be approximately measured directly by measuring the average power that remains in a signal after all the frequencies outside a narrow band have been filtered out.

Spectral analysis is carried out for visual signals as well. The power spectrum ignores all phase relations, which is good enough for many purposes, but for video signals other types of spectral analysis must also be employed, still using the Fourier transform as a tool.

2.12 Other notations

Other common notations for $\hat{f}(\xi)$ include:

$$\tilde{f}(\xi), \ \tilde{f}(\omega), \ F(\xi), \ \mathcal{F}(f)(\xi), \ (\mathcal{F}f)(\xi), \ \mathcal{F}(f), \ \mathcal{F}(\omega), \ F(\omega), \ \mathcal{F}(j\omega), \ \mathcal{F}\{f\}, \ \mathcal{F}(f(t)), \ \mathcal{F}\{f(t)\}.$$

Denoting the Fourier transform by a capital letter corresponding to the letter of function being transformed (such as $f(x)$ and $F(\xi)$) is especially common in the sciences and engineering. In electronics, the omega (ω) is often used instead of ξ due to its interpretation as angular frequency, sometimes it is written as $F(j\omega)$, where j is the imaginary unit, to indicate its relationship with the Laplace transform, and sometimes it is written informally as $F(2\pi f)$ in order to use ordinary frequency.

The interpretation of the complex function $\hat{f}(\xi)$ may be aided by expressing it in polar coordinate form

$$\hat{f}(\xi) = A(\xi) e^{i\varphi(\xi)}$$

in terms of the two real functions $A(\xi)$ and $\varphi(\xi)$ where:

$$A(\xi) = |\hat{f}(\xi)|,$$

is the amplitude and

$$\varphi(\xi) = \arg\left(\hat{f}(\xi)\right),$$

is the phase (see arg function).

Then the inverse transform can be written:

$$f(x) = \int_{-\infty}^{\infty} A(\xi)\, e^{i(2\pi\xi x + \varphi(\xi))}\, d\xi,$$

which is a recombination of all the **frequency components** of $f(x)$. Each component is a complex sinusoid of the form $e^{2\pi ix\xi}$ whose amplitude is $A(\xi)$ and whose initial phase angle (at $x = 0$) is $\varphi(\xi)$.

The Fourier transform may be thought of as a mapping on function spaces. This mapping is here denoted $\mathcal{F}$ and $\mathcal{F}(f)$ is used to denote the Fourier transform of the function f. This mapping is linear, which means that $\mathcal{F}$ can also be seen as a linear transformation on the function space and implies that the standard notation in linear algebra of applying a linear transformation to a vector (here the function f) can be used to write $\mathcal{F}f$ instead of $\mathcal{F}(f)$. Since the result of applying the Fourier transform is again a function, we can be interested in the value of this function evaluated at the value ξ for its variable, and this is denoted either as $\mathcal{F}f(\xi)$ or as $(\mathcal{F}f)(\xi)$. Notice that in the former case, it is implicitly understood that $\mathcal{F}$ is applied first to f and then the resulting function is evaluated at ξ, not the other way around.

In mathematics and various applied sciences, it is often necessary to distinguish between a function f and the value of f when its variable equals x, denoted $f(x)$. This means that a notation like $\mathcal{F}(f(x))$ formally can be interpreted as the Fourier transform of the values of f at x. Despite this flaw, the previous notation appears frequently, often when a particular function or a function of a particular variable is to be transformed.

For example, $\mathcal{F}(\text{rect}(x)) = \text{sinc}(\xi)$ is sometimes used to express that the Fourier transform of a rectangular function is a sinc function,

or $\mathcal{F}(f(x + x_0)) = \mathcal{F}(f(x))e^{2\pi i\xi x_0}$ is used to express the shift property of the Fourier transform.

Notice, that the last example is only correct under the assumption that the transformed function is a function of x, not of x_0.

2.13   Other conventions

The Fourier transform can also be written in terms of angular frequency:

$$\omega = 2\pi\xi,$$

whose units are radians per second.

The substitution $\xi = \omega/(2\pi)$ into the formulas above produces this convention:

$$\hat{f}(\omega) = \int_{\mathbf{R}^n} f(x)e^{-i\omega \cdot x}\, dx.$$

Under this convention, the inverse transform becomes:

$$f(x) = \frac{1}{(2\pi)^n} \int_{\mathbf{R}^n} \hat{f}(\omega)e^{i\omega \cdot x}\, d\omega.$$

Unlike the convention followed in this article, when the Fourier transform is defined this way, it is no longer a unitary transformation on $L^2(\mathbf{R}^n)$. There is also less symmetry between the formulas for the Fourier transform and its inverse.

Another convention is to split the factor of $(2\pi)^n$ evenly between the Fourier transform and its inverse, which leads to definitions:

$$\hat{f}(\omega) = \frac{1}{(2\pi)^{n/2}} \int_{\mathbf{R}^n} f(x)e^{-i\omega \cdot x}\, dx,$$

$$f(x) = \frac{1}{(2\pi)^{n/2}} \int_{\mathbf{R}^n} \hat{f}(\omega) e^{i\omega \cdot x} \, d\omega.$$

Under this convention, the Fourier transform is again a unitary transformation on $L^2(\mathbf{R}^n)$. It also restores the symmetry between the Fourier transform and its inverse.

Variations of all three conventions can be created by conjugating the complex-exponential kernel of both the forward and the reverse transform. The signs must be opposites. Other than that, the choice is (again) a matter of convention.

As discussed above, the characteristic function of a random variable is the same as the Fourier–Stieltjes transform of its distribution measure, but in this context it is typical to take a different convention for the constants. Typically characteristic function is defined $E(e^{it \cdot X}) = \int e^{it \cdot x} \, d\mu_X(x)$.

As in the case of the "non-unitary angular frequency" convention above, there is no factor of 2π appearing in either of the integral, or in the exponential. Unlike any of the conventions appearing above, this convention takes the opposite sign in the exponential.

2.14 Computation Methods

The appropriate computation method largely depends how the original mathematical function is represented and the desired form of the output function.

Since the fundamental definition of a Fourier transform is an integral, functions that can be expressed as closed-form expressions are commonly computed by working the integral analytically to yield a closed-form expression in the Fourier transform conjugate variable as the result. This is the method used to generate tables of Fourier transforms,[18] including those found in the table below (Fourier transform#Tables of important Fourier transforms).

Many computer algebra systems such as Matlab and Mathematica that are capable of symbolic integration are capable of computing Fourier transforms analytically. For example, to compute the Fourier transform of $f(t) = \cos(6\pi t) \, e^{-\pi t^2}$ one might enter the command "integrate cos(6*pi*t) exp(−pi*t^2) exp(-i*2*pi*f*t) from -inf to inf" into Wolfram Alpha.

2.14.1 Numerical integration of closed-form functions

If the input function is in closed-form and the desired output function is a series of ordered pairs (for example a table of values from which a graph can be generated) over a specified domain, then the Fourier transform can be generated by numerical integration at each value of the Fourier conjugate variable (frequency, for example) for which a value of the output variable is desired.[19] Note that this method requires computing a separate numerical integration for each value of frequency for which a value of the Fourier transform is desired.[20][21] The numerical integration approach works on a much broader class of functions than the analytic approach, because it yields resuls for functions that do not have closed form Fourier transform integrals.

2.14.2 Numerical integration of a series of ordered pairs

If the input function is a series of ordered pairs (for example, a time series from measuring an output variable repeatedly over a time interval) then the output function must also be a series of ordered pairs (for example, a complex number vs. frequency over a specified domain of frequencies), unless certain assumptions and approximations are made allowing the output function to be approximated by a closed-form expression. In the general case where the available input series of ordered pairs are assumed be samples representing a continuous function over an interval (amplitude vs. time, for example), the series of ordered pairs representing the desired output function can be obtained by numerical integration of the input data over the available interval at each value of the Fourier conjugate variable (frequency, for example) for which the value of the Fourier transform is desired.[22]

Explicit numerical integration over the ordered pairs can yield the Fourier transform output value for any desired value of the conjugate Fourier transform variable (frequency, for example), so that a spectrum can be produced at any desired step size and over any desired variable range for accurate determination of amplitudes, frequencies, and phases corresponding to isolated peaks. Unlike limitations in DFT and FFT methods, explicit numerical integration can have any desired step size and compute the Fourier transform over any desired range of the conjugate Fourier transform variable (for example, frequency).

2.14.3   Discrete Fourier Transforms and Fast Fourier Transforms

If the ordered pairs representing the original input function are equally spaced in their input variable (for example, equal time steps), then the Fourier transform is known as a discrete Fourier transform (DFT), which can be computed either by explicit numerical integration, by explicit evaluation of the DFT definition, or by fast Fourier transform (FFT) methods. In contrast to explicit integration of input data, use of the DFT and FFT methods produces Fourier transforms described by ordered pairs of step size equal to the reciprocal of the original sampling interval. For example, if the input data is sampled for 10 seconds, the output of DFT and FFT methods will have a 0.1 Hz frequency spacing.

2.15   Tables of important Fourier transforms

The following tables record some closed-form Fourier transforms. For functions $f(x)$, $g(x)$ and $h(x)$ denote their Fourier transforms by $\hat{f}$, $\hat{g}$, and $\hat{h}$ respectively. Only the three most common conventions are included. It may be useful to notice that entry 105 gives a relationship between the Fourier transform of a function and the original function, which can be seen as relating the Fourier transform and its inverse.

2.15.1   Functional relationships

The Fourier transforms in this table may be found in Erdélyi (1954) or Kammler (2000, appendix).

2.15.2   Square-integrable functions

The Fourier transforms in this table may be found in (Campbell & Foster 1948), (Erdélyi 1954), or the appendix of (Kammler 2000).

2.15.3   Distributions

The Fourier transforms in this table may be found in (Erdélyi 1954) or the appendix of (Kammler 2000).

2.15.4   Two-dimensional functions

Remarks

To 400: The variables ξx, ξy, ωx, ωy, $v x$ and $v y$ are real numbers. The integrals are taken over the entire plane.

To 401: Both functions are Gaussians, which may not have unit volume.

To 402: The function is defined by $\operatorname{circ}(r) = 1\ 0 \le r \le 1$, and is 0 otherwise. This is the Airy distribution, and is expressed using J_1 (the order 1 Bessel function of the first kind). (Stein & Weiss 1971, Thm. IV.3.3)

2.15.5   Formulas for general *n*-dimensional functions

Remarks

To 501: The function $\chi_{[0,\,1]}$ is the indicator function of the interval $[0, 1]$. The function $\Gamma(x)$ is the gamma function. The function $J_{n/2\,+\,\delta}$ is a Bessel function of the first kind, with order $n/2 + \delta$. Taking $n = 2$ and $\delta = 0$ produces 402. (Stein & Weiss 1971, Thm. 4.15)

To 502: See Riesz potential where the constant is given by $c_{\alpha,n} = \pi^{n/2} 2^{\alpha} \frac{\Gamma(\alpha/2)}{\Gamma((n-\alpha)/2)}$. The formula also holds for all $\alpha \neq -n, -n-1, \ldots$ by analytic continuation, but then the function and its Fourier transforms need to be understood as suitably regularized tempered distributions. See homogeneous distribution.

To 503: This is the formula for a multivariate normal distribution normalized to 1 with a mean of 0. Bold variables are vectors or matrices. Following the notation of the aforementioned page, $\boldsymbol{\Sigma} = \boldsymbol{\sigma}\boldsymbol{\sigma}^{\mathrm{T}}$ and $\boldsymbol{\Sigma}^{-1} = \boldsymbol{\sigma}^{-\mathrm{T}}\boldsymbol{\sigma}^{-1}$

To 504: Here $c_n = \Gamma((n+1)/2)/\pi^{(n+1)/2}$. See (Stein & Weiss 1971, p. 6).

2.16   See also

- Analog signal processing

- Beevers–Lipson strip

- Discrete Fourier transform

    - DFT matrix

- Discrete-time Fourier transform

- Fast Fourier transform

- Fourier integral operator

- Fourier inversion theorem

- Fourier multiplier

- Fourier series

- Fourier sine transform

- Fourier–Deligne transform

- Fourier–Mukai transform

- Fractional Fourier transform

- Indirect Fourier transform

- Integral transform

    - Hankel transform
    - Hartley transform

- Laplace transform

- Linear canonical transform

- Mellin transform

- Multidimensional transform

- NGC 4622, especially the image NGC 4622 Fourier transform $m = 2$.

- Short-time Fourier transform

- Space-time Fourier transform

- Spectral density

  - Spectral density estimation

- Symbolic integration

- Time stretch dispersive Fourier transform

- Transform (mathematics)

2.17   Remarks

[1] Up to an imaginary constant factor whose magnitude depends on what Fourier transform convention is used.

[2] The Laplace transform is a generalization of the Fourier transform that offers greater flexibility for many such applications.

[3] Depending on the application a Lebesgue integral, distributional, or other approach may be most appropriate.

2.18   Notes

[1] Vretblad 2000 provides solid justification for these formal procedures without going too deeply into functional analysis or the theory of distributions.

[2] In relativistic quantum mechanics one encounters vector-valued Fourier transforms of multi-component wave functions. In quantum field theory operator-valued Fourier transforms of operator-valued functions of spacetime are in frequent use, see for example Greiner & Reinhardt 1996.

[3] Fourier, Joseph (1822). *Théorie analytique de la chaleur* (in French). Paris: Firmin Didot Père et Fils. OCLC 2688081.

[4] Paley & Wiener 1934

[5] Guelfand & Vilenkin 1967, p. 19

[6] Kirillov & Gvichiani 1982, pp. 114,226

[7] Clozel & Delorme 1985,, pp. 331–333

[8] de Groot & Mazur 1987, p. 146

[9] Champeney 1987, p. 80

[10] Chatfield 2004, p. 297

[11] Kolmogórov & Fomín 1978, p. 492

[12] Wiener 1949

[13] Kolmogórov & Fomín 1978, p. 492

[14] Champeney 1987, p. 63

[15] Widder & Wiener 1938, p. 537

[16] hatfield, The Analysis of Time Series, 6th ed., London, 2004, p. 113.

[17] See, e.g., Fourier, Théorie Analytique de la Chaleur, Paris, 1822, p. 441; Poincaré, Théorie Analytique de la Propagation de Chaleur, Paris, 1895, p. 102; Whittaker and Watson, A Course of Modern Analysis, 4th ed., Cambridge, 1927, p. 188), or "Fourier's Integral Theorem" (although it was not proved in this generality by Fourier) and is equivalent to "Fourier Inversion". See Camille Jordan, Cours d'Analyse de l'École Polytechnique, vol. II, Calcul Intégral: Intégrales définies et indéfinies. 2nd ed., Paris, 1883, pp. 216–226, who, in fact, proves Fourier's Integral Theorem before studying Fourier Series. See also Kolmogorov and Fomin, Elementos de la teoría de funciones y del análisis funcional. Moscow, 1972, traducido por Carlos Vega, pp. 466–9.

[18] Gradshteyn and Ryzhik's Table of Integrals, Series, and Products Daniel Zwillinger and Victor Moll (eds.) Eighth edition (Oct 2014)

[19] Press, William H., et al. Numerical recipes in C. Vol. 2. Cambridge: Cambridge university press, 1996.

[20] Bailey, David H., and Paul N. Swarztrauber. "A fast method for the numerical evaluation of continuous Fourier and Laplace transforms." SIAM Journal on Scientific Computing 15.5 (1994): 1105–1110.

[21] Lado, F. "Numerical fourier transforms in one, two, and three dimensions for liquid state calculations." Journal of Computational Physics 8.3 (1971): 417–433.

[22] Simonen, P., and H. Olkkonen. "Fast method for computing the Fourier integral transform via Simpson's numerical integration." Journal of biomedical engineering 7.4 (1985): 337–340.

2.19 References

- Boashash, B., ed. (2003), *Time-Frequency Signal Analysis and Processing: A Comprehensive Reference*, Oxford: Elsevier Science, ISBN 0-08-044335-4

- Bochner S., Chandrasekharan K. (1949), *Fourier Transforms*, Princeton University Press

- Bracewell, R. N. (2000), *The Fourier Transform and Its Applications* (3rd ed.), Boston: McGraw-Hill, ISBN 0-07-116043-4.

- Campbell, George; Foster, Ronald (1948), *Fourier Integrals for Practical Applications*, New York: D. Van Nostrand Company, Inc..

- Condon, E. U. (1937), "Immersion of the Fourier transform in a continuous group of functional transformations", *Proc. Nat. Acad. Sci. USA* **23**: 158–164.

- Duoandikoetxea, Javier (2001), *Fourier Analysis*, American Mathematical Society, ISBN 0-8218-2172-5.

- Dym, H; McKean, H (1985), *Fourier Series and Integrals*, Academic Press, ISBN 978-0-12-226451-1.

- Erdélyi, Arthur, ed. (1954), *Tables of Integral Transforms* **1**, New Your: McGraw-Hill

- Fourier, J. B. Joseph (1822), *Théorie Analytique de la Chaleur*, Paris: Chez Firmin Didot, père et fils

- Fourier, J. B. Joseph; Freeman, Alexander, translator (1878), *The Analytical Theory of Heat*, The University Press

- Champeney, D.C. (1987), *A Handbook of Fourier Theorems*, Cambridge University Press

- de Groot, Sybren R.; Mazur, Peter (1984), *Non-Equilibrium Thermodynamics* (2nd ed.), New York: Dover

- Marín Antuña, José (1990), *Teoría de funciones de variable compleja* (2nd ed.), Havana: Editorial Pueblo y Educación

- Chatfield, Chris (2004), *The Analysis of Time Series: An Introduction*, Texts in Statistical Science (6th ed.), London: Chapman & Hall/CRC

- Feller, William (1971), *An Introduction to Probability Theory and Its Applications. Vol. II.* (Second ed.), New York: John Wiley & Sons, MR 0270403.

- Wiener, Norbert (1949), *Extrapolation, Interpolation, and Smoothing of Stationary Time Series With Engineering Applications*, Cambridge, Mass.: Technology Press and John Wiley & Sons and Chapman & Hall

- Kirillov, Alexandre; Gvichiani, Alexei (1982), *Théorèmes et problèmes d'analyse fonctionnelle*, Djilali Embarek, translator, Moscow: Mir

- Kolmogórov, Andréi Nikolaevich; Fomín, Serguei Vasílievich (1978), *Elementos de la teoría de funciones y del análisis functional*, Carlos Vega, translator (3rd ed.), Moscow: Mir

- Guelfand, Israel Moiseevich; Vilenkin, N.Y. (1967), *Les distributions tome 4: applications de l'analyse harmonique*, G. Rideau, translator, Paris: Dunod

- Widder, David Vernon; Wiener, Norbert (August 1938), "Remarks on the Classical Inversion Formula for the Laplace Integral", *Bulletin of the American Mathematical Society* **44**: 573–575, doi:10.1090/s0002-9904-1938-06812-7

- Clozel, Laurent; Delorme, Patrice (1985), "Sur le théorème de Paley-Wiener invariant pour les groupes de Lie réductifs réels", *C. R. Acad. Sci. Paris, série I* **300**: 331–333

- Paley, R.E.A.C.; Wiener, Norbert (1934), *Fourier Transforms in the Complex Domain*, American Mathematical Society Colloquium Publications (19), Providence, Rhode Island: American Mathematical Society

- Grafakos, Loukas (2004), *Classical and Modern Fourier Analysis*, Prentice-Hall, ISBN 0-13-035399-X.

- Grafakos, Loukas; Teschl, Gerald (2013), "On Fourier transforms of radial functions and distributions", *J. Fourier Anal. Appl.* **19**: 167–179, doi:10.1007/s00041-012-9242-5.

- Greiner, W.; Reinhardt, J. (1996), *Field Quantization*, Springer Publishing, ISBN 3-540-59179-6

- Hewitt, Edwin; Ross, Kenneth A. (1970), *Abstract harmonic analysis. Vol. II: Structure and analysis for compact groups. Analysis on locally compact Abelian groups*, Die Grundlehren der mathematischen Wissenschaften, Band 152, Berlin, New York: Springer-Verlag, MR 0262773.

- Hörmander, L. (1976), *Linear Partial Differential Operators, Volume 1*, Springer-Verlag, ISBN 978-3-540-00662-6.

- James, J.F. (2011), *A Student's Guide to Fourier Transforms* (3rd ed.), New York: Cambridge University Press, ISBN 978-0-521-17683-5.

- Kaiser, Gerald (1994), *A Friendly Guide to Wavelets*, Birkhäuser, ISBN 0-8176-3711-7

- Kammler, David (2000), *A First Course in Fourier Analysis*, Prentice Hall, ISBN 0-13-578782-3

- Katznelson, Yitzhak (1976), *An introduction to Harmonic Analysis*, Dover, ISBN 0-486-63331-4

- Knapp, Anthony W. (2001), *Representation Theory of Semisimple Groups: An Overview Based on Examples*, Princeton University Press, ISBN 978-0-691-09089-4

- Pinsky, Mark (2002), *Introduction to Fourier Analysis and Wavelets*, Brooks/Cole, ISBN 0-534-37660-6

- Polyanin, A. D.; Manzhirov, A. V. (1998), *Handbook of Integral Equations*, Boca Raton: CRC Press, ISBN 0-8493-2876-4.

- Rudin, Walter (1987), *Real and Complex Analysis* (Third ed.), Singapore: McGraw Hill, ISBN 0-07-100276-6.

- Rahman, Matiur (2011), *Applications of Fourier Transforms to Generalized Functions*, WIT Press, ISBN 1845645642.

- Stein, Elias; Shakarchi, Rami (2003), *Fourier Analysis: An introduction*, Princeton University Press, ISBN 0-691-11384-X.

- Stein, Elias; Weiss, Guido (1971), *Introduction to Fourier Analysis on Euclidean Spaces*, Princeton, N.J.: Princeton University Press, ISBN 978-0-691-08078-9.

- Taneja, HC (2008), "Chapter 18: Fourier integrals and Fourier transforms", *Advanced Engineering Mathematics:, Volume 2*, New Delhi, India: I. K. International Pvt Ltd, ISBN 8189866567.

- Titchmarsh, E (1948), *Introduction to the theory of Fourier integrals* (2nd ed.), Oxford University: Clarendon Press (published 1986), ISBN 978-0-8284-0324-5.

- Vretblad, Anders (2000), *Fourier Analysis and its Applications*, Graduate Texts in Mathematics **223**, New York: Springer Publishing, ISBN 0-387-00836-5

- Wilson, R. G. (1995), *Fourier Series and Optical Transform Techniques in Contemporary Optics*, New York: Wiley, ISBN 0-471-30357-7.

- Yosida, K. (1968), *Functional Analysis*, Springer-Verlag, ISBN 3-540-58654-7.

2.20 External links

- Weisstein, Eric W., "Fourier Transform", *MathWorld*.

Chapter 3

Fourier series

In mathematics, a **Fourier series** (English pronunciation: /ˈfɔəriɛɪ/) is a way to represent a (wave-like) function as the sum of simple sine waves. More formally, it decomposes any periodic function or periodic signal into the sum of a (possibly infinite) set of simple oscillating functions, namely sines and cosines (or, equivalently, complex exponentials). The discrete-time Fourier transform is a periodic function, often defined in terms of a Fourier series. The Z-transform, another example of application, reduces to a Fourier series for the important case $|z|=1$. Fourier series are also central to the original proof of the Nyquist–Shannon sampling theorem. The study of Fourier series is a branch of Fourier analysis.

3.1 History

See also: Fourier analysis § History

The Fourier series is named in honour of Jean-Baptiste Joseph Fourier (1768–1830), who made important contributions to the study of trigonometric series, after preliminary investigations by Leonhard Euler, Jean le Rond d'Alembert, and Daniel Bernoulli.[nb 1] Fourier introduced the series for the purpose of solving the heat equation in a metal plate, publishing his initial results in his 1807 *Mémoire sur la propagation de la chaleur dans les corps solides* (*Treatise on the propagation of heat in solid bodies*), and publishing his *Théorie analytique de la chaleur* (*Analytical theory of heat*) in 1822. Early ideas of decomposing a periodic function into the sum of simple oscillating functions date back to the 3rd century BC, when ancient astronomers proposed an empiric model of planetary motions, based on deferents and epicycles.

The heat equation is a partial differential equation. Prior to Fourier's work, no solution to the heat equation was known in the general case, although particular solutions were known if the heat source behaved in a simple way, in particular, if the heat source was a sine or cosine wave. These simple solutions are now sometimes called eigensolutions. Fourier's idea was to model a complicated heat source as a superposition (or linear combination) of simple sine and cosine waves, and to write the solution as a superposition of the corresponding eigensolutions. This superposition or linear combination is called the Fourier series.

From a modern point of view, Fourier's results are somewhat informal, due to the lack of a precise notion of function and integral in the early nineteenth century. Later, Peter Gustav Lejeune Dirichlet[1] and Bernhard Riemann[2][3][4] expressed Fourier's results with greater precision and formality.

Although the original motivation was to solve the heat equation, it later became obvious that the same techniques could be applied to a wide array of mathematical and physical problems, and especially those involving linear differential equations with constant coefficients, for which the eigensolutions are sinusoids. The Fourier series has many such applications in electrical engineering, vibration analysis, acoustics, optics, signal processing, image processing, quantum mechanics, econometrics,[5] thin-walled shell theory,[6] etc.

3.2 Definition

In this section, $s(x)$ denotes a function of the real variable x, and s is integrable on an interval $[x_0, x_0 + P]$, for real numbers x_0 and P. We will attempt to represent s in that interval as an infinite sum, or series, of harmonically related sinusoidal functions. Outside the interval, the series is periodic with period P (frequency $1/P$). It follows that if s also has that property, the approximation is valid on the entire real line. We can begin with a finite summation (or *partial sum*):

$$s_N(x) = \frac{A_0}{2} + \sum_{n=1}^{N} A_n \cdot \sin\left(\frac{2\pi nx}{P} + \phi_n\right), \quad \text{integer for } N \geq 1.$$

$s_N(x)$ is a periodic function with period **P**. Using the identities:

$$\sin\left(\frac{2\pi nx}{P} + \phi_n\right) \equiv \sin(\phi_n)\cos\left(\frac{2\pi nx}{P}\right) + \cos(\phi_n)\sin\left(\frac{2\pi nx}{P}\right)$$

$$\sin\left(\frac{2\pi nx}{P} + \phi_n\right) \equiv \mathrm{Re}\left\{\frac{1}{i} \cdot e^{i\left(\frac{2\pi nx}{P} + \phi_n\right)}\right\} = \frac{1}{2i} \cdot e^{i\left(\frac{2\pi nx}{P} + \phi_n\right)} + \left(\frac{1}{2i} \cdot e^{i\left(\frac{2\pi nx}{P} + \phi_n\right)}\right)^{*},$$

we can also write the function in these equivalent forms:

The value of the coefficient A_n and the phaseshift ϕ_n can be calculated with the following formula:

$$A_n = \sqrt{a_n{}^2 + b_n{}^2}$$

$$\phi_n = \arctan\left(\frac{a_n}{b_n}\right)$$

based on the Fourier coefficients as calculated below.

where:

$$c_n \overset{\text{def}}{=} \begin{cases} \frac{A_n}{2i} e^{i\phi_n} = \frac{1}{2}(a_n - ib_n) & \text{for } n > 0 \\ \frac{1}{2}a_0 & \text{for } n = 0 \\ c_{|n|}^{*} & \text{for } n < 0. \end{cases}$$

When the coefficients (known as **Fourier coefficients**) are computed as follows:[7]

$s_N(x)$ approximates $s(x)$ on $[x_0, x_0+P]$, and the approximation improves as $N \to \infty$. The infinite sum, $s_\infty(x)$, is called the **Fourier series** representation of s. In engineering applications, the Fourier series is generally presumed to converge everywhere except at discontinuities, since the functions encountered in engineering are more well behaved than the ones that mathematicians can provide as counter-examples to this presumption. In particular, the Fourier series converges absolutely and uniformly to $s(x)$ whenever the derivative of $s(x)$ (which may not exist everywhere) is square integrable.[8] If a function is square-integrable on the interval $[x_0, x_0+P]$, then the Fourier series converges to the function at *almost every* point. Convergence of Fourier series also depends on the finite number of maxima and minima in a function which is popularly known as one of the Dirichlet's condition for Fourier series. See Convergence of Fourier series. It is possible to define Fourier coefficients for more general functions or distributions, in such cases convergence in norm or weak convergence is usually of interest.

- Another visualisation of an approximation of a square wave by taking the first 1, 2, 3 and 4 terms of its Fourier series. (An interactive animation can be seen here)

- A visualisation of an approximation of a sawtooth wave of the same amplitude and frequency for comparison

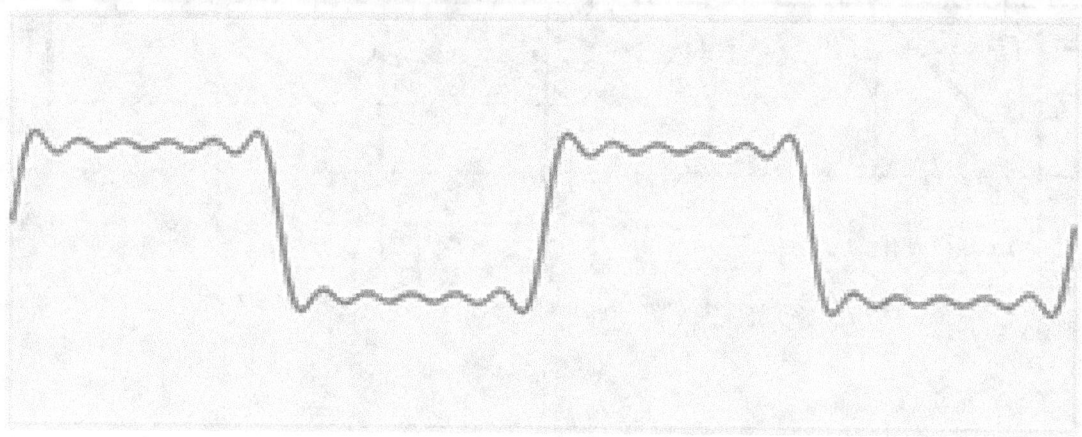

Function s(x) (in red) is a sum of six sine functions of different amplitudes and harmonically related frequencies. Their summation is called a Fourier series. The Fourier transform, S(f) (in blue), which depicts amplitude vs frequency, reveals the 6 frequencies and their amplitudes.

3.2.1   Example 1: a simple Fourier series

We now use the formula above to give a Fourier series expansion of a very simple function. Consider a sawtooth wave

$$s(x) = \frac{x}{\pi}, \quad \text{for} - \pi < x < \pi,$$

$$s(x + 2\pi k) = s(x), \quad \text{for} - \infty < x < \infty \text{ and } k \in \mathbb{Z}.$$

In this case, the Fourier coefficients are given by

$$a_n = \frac{1}{\pi} \int_{-\pi}^{\pi} s(x) \cos(nx)\, dx = 0, \quad n \geq 0.$$

$$b_n = \frac{1}{\pi} \int_{-\pi}^{\pi} s(x) \sin(nx)\, dx$$

$$= -\frac{2}{\pi n} \cos(n\pi) + \frac{2}{\pi^2 n^2} \sin(n\pi)$$

$$= \frac{2\,(-1)^{n+1}}{\pi n}, \quad n \geq 1.$$

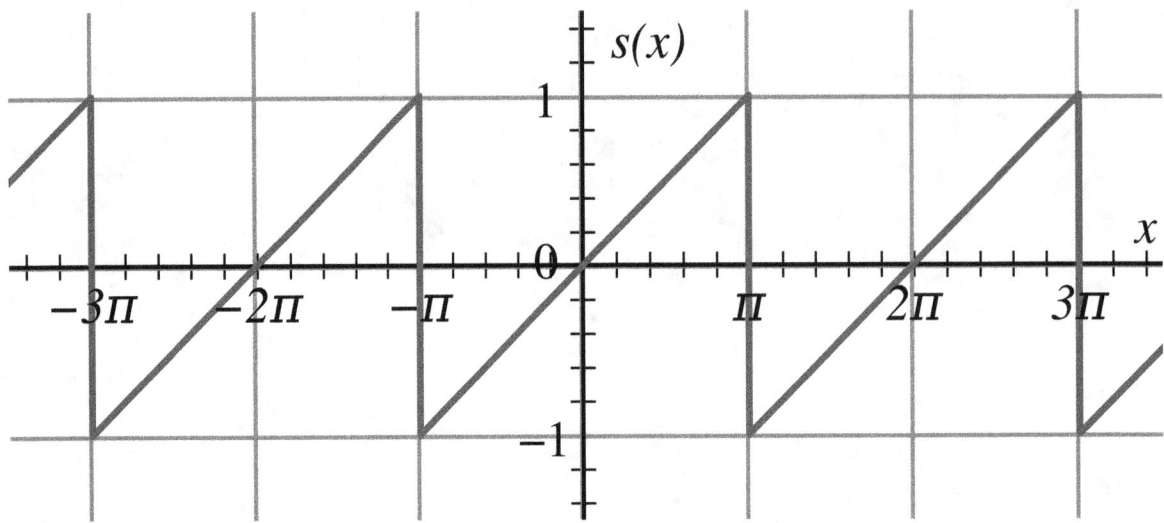

Plot of the sawtooth wave, a periodic continuation of the linear function $s(x) = x/\pi$ *on the interval* $(-\pi, \pi]$

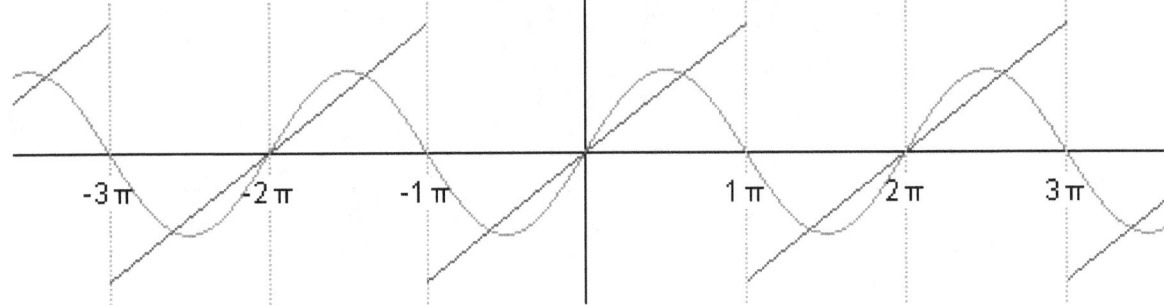

Animated plot of the first five successive partial Fourier series

It can be proven that Fourier series converges to $s(x)$ at every point x where s is differentiable, and therefore:

When $x = \pi$, the Fourier series converges to 0, which is the half-sum of the left- and right-limit of s at $x = \pi$. This is a particular instance of the Dirichlet theorem for Fourier series.

This example leads us to a solution to the Basel problem.

3.2.2 Example 2: Fourier's motivation

The Fourier series expansion of our function in Example 1 looks more complicated than the simple formula $s(x) = x/\pi$, so it is not immediately apparent why one would need the Fourier series. While there are many applications, Fourier's motivation was in solving the heat equation. For example, consider a metal plate in the shape of a square whose side measures π meters, with coordinates $(x, y) \in [0, \pi] \times [0, \pi]$. If there is no heat source within the plate, and if three of the four sides are held at 0 degrees Celsius, while the fourth side, given by $y = \pi$, is maintained at the temperature gradient $T(x, \pi) = x$ degrees Celsius, for x in $(0, \pi)$, then one can show that the stationary heat distribution (or the heat distribution after a long period of time has elapsed) is given by

$$T(x,y) = 2\sum_{n=1}^{\infty} \frac{(-1)^{n+1}}{n} \sin(nx)\frac{\sinh(ny)}{\sinh(n\pi)}.$$

Heat distribution in a metal plate, using Fourier's method

Here, sinh is the hyperbolic sine function. This solution of the heat equation is obtained by multiplying each term of **Eq.1** by $\sinh(ny)/\sinh(n\pi)$. While our example function $s(x)$ seems to have a needlessly complicated Fourier series, the heat distribution $T(x, y)$ is nontrivial. The function T cannot be written as a closed-form expression. This method of solving the heat problem was made possible by Fourier's work.

3.2.3   Other applications

Another application of this Fourier series is to solve the Basel problem by using Parseval's theorem. The example generalizes and one may compute $\zeta(2n)$, for any positive integer n.

3.2.4 Other common notations

The notation *cn* is inadequate for discussing the Fourier coefficients of several different functions. Therefore, it is customarily replaced by a modified form of the function (*s*, in this case), such as $\hat{s}$ or S, and functional notation often replaces subscripting:

$$s_\infty(x) = \sum_{n=-\infty}^{\infty} \hat{s}(n) \cdot e^{i\frac{2\pi nx}{P}}$$

$$= \sum_{n=-\infty}^{\infty} S[n] \cdot e^{j\frac{2\pi nx}{P}} \quad \text{\scriptsize notation engineering common}$$

In engineering, particularly when the variable *x* represents time, the coefficient sequence is called a frequency domain representation. Square brackets are often used to emphasize that the domain of this function is a discrete set of frequencies.

Another commonly used frequency domain representation uses the Fourier series coefficients to modulate a Dirac comb:

$$S(f) \stackrel{\text{def}}{=} \sum_{n=-\infty}^{\infty} S[n] \cdot \delta\left(f - \frac{n}{P}\right),$$

where *f* represents a continuous frequency domain. When variable *x* has units of seconds, *f* has units of hertz. The "teeth" of the comb are spaced at multiples (i.e. harmonics) of 1/P, which is called the fundamental frequency. $s_\infty(x)$ can be recovered from this representation by an inverse Fourier transform:

$$\mathcal{F}^{-1}\{S(f)\} = \int_{-\infty}^{\infty} \left(\sum_{n=-\infty}^{\infty} S[n] \cdot \delta\left(f - \frac{n}{P}\right) \right) e^{i2\pi fx} \, df,$$

$$= \sum_{n=-\infty}^{\infty} S[n] \cdot \int_{-\infty}^{\infty} \delta\left(f - \frac{n}{P}\right) e^{i2\pi fx} \, df,$$

$$= \sum_{n=-\infty}^{\infty} S[n] \cdot e^{i\frac{2\pi nx}{P}} \stackrel{\text{def}}{=} s_\infty(x).$$

The constructed function $S(f)$ is therefore commonly referred to as a **Fourier transform**, even though the Fourier integral of a periodic function is not convergent at the harmonic frequencies.[nb 2]

3.3 Beginnings

This immediately gives any coefficient *ak* of the trigonometrical series for $\varphi(y)$ for any function which has such an expansion. It works because if φ has such an expansion, then (under suitable convergence assumptions) the integral

$$a_k = \int_{-1}^{1} \varphi(y) \cos(2k+1)\frac{\pi y}{2} \, dy$$

$$= \int_{-1}^{1} \left(a \cos\frac{\pi y}{2} \cos(2k+1)\frac{\pi y}{2} + a' \cos 3\frac{\pi y}{2} \cos(2k+1)\frac{\pi y}{2} + \cdots \right) dy$$

can be carried out term-by-term. But all terms involving $\cos(2j+1)\frac{\pi y}{2} \cos(2k+1)\frac{\pi y}{2}$ for $j \neq k$ vanish when integrated from −1 to 1, leaving only the *k*th term.

In these few lines, which are close to the modern formalism used in Fourier series, Fourier revolutionized both mathematics and physics. Although similar trigonometric series were previously used by Euler, d'Alembert, Daniel Bernoulli and Gauss, Fourier believed that such trigonometric series could represent any arbitrary function. In what sense that is actually true is a somewhat subtle issue and the attempts over many years to clarify this idea have led to important discoveries in the theories of convergence, function spaces, and harmonic analysis.

When Fourier submitted a later competition essay in 1811, the committee (which included Lagrange, Laplace, Malus and Legendre, among others) concluded: *...the manner in which the author arrives at these equations is not exempt of difficulties and...his analysis to integrate them still leaves something to be desired on the score of generality and even rigour.*

3.3.1   Birth of harmonic analysis

Since Fourier's time, many different approaches to defining and understanding the concept of Fourier series have been discovered, all of which are consistent with one another, but each of which emphasizes different aspects of the topic. Some of the more powerful and elegant approaches are based on mathematical ideas and tools that were not available at the time Fourier completed his original work. Fourier originally defined the Fourier series for real-valued functions of real arguments, and using the sine and cosine functions as the basis set for the decomposition.

Many other Fourier-related transforms have since been defined, extending the initial idea to other applications. This general area of inquiry is now sometimes called harmonic analysis. A Fourier series, however, can be used only for periodic functions, or for functions on a bounded (compact) interval.

3.4   Extensions

3.4.1   Fourier series on a square

We can also define the Fourier series for functions of two variables x and y in the square $[-\pi, \pi] \times [-\pi, \pi]$:

$$f(x,y) = \sum_{j,k \in \mathbf{Z}(\text{integers})} c_{j,k} e^{ijx} e^{iky},$$

$$c_{j,k} = \frac{1}{4\pi^2} \int_{-\pi}^{\pi} \int_{-\pi}^{\pi} f(x,y) e^{-ijx} e^{-iky} \, dx \, dy.$$

Aside from being useful for solving partial differential equations such as the heat equation, one notable application of Fourier series on the square is in image compression. In particular, the jpeg image compression standard uses the two-dimensional discrete cosine transform, which is a Fourier transform using the cosine basis functions.

3.4.2   Fourier series of Bravais-lattice-periodic-function

The Bravais lattice is defined as the set of vectors of the form:

$$\mathbf{R} = n_1 \mathbf{a}_1 + n_2 \mathbf{a}_2 + n_3 \mathbf{a}_3$$

where n_i are integers and $\mathbf{a}_i$ are three linearly independent vectors. Assuming we have some function, $f(\mathbf{r})$, such that it obeys the following condition for any Bravais lattice vector $\mathbf{R}$: $f(\mathbf{r}) = f(\mathbf{r} + \mathbf{R})$, we could make a Fourier series of it. This kind of function can be, for example, the effective potential that one electron "feels" inside a periodic crystal. It is useful to make a Fourier series of the potential then when applying Bloch's theorem. First, we may write any arbitrary vector $\mathbf{r}$ in the coordinate-system of the lattice:

$$\mathbf{r} = x_1 \frac{\mathbf{a_1}}{a_1} + x_2 \frac{\mathbf{a_2}}{a_2} + x_3 \frac{\mathbf{a_3}}{a_3},$$

where $ai = |\mathbf{a}i|$.

Thus we can define a new function,

$$g(x_1, x_2, x_3) := f(\mathbf{r}) = f\left(x_1 \frac{\mathbf{a_1}}{a_1} + x_2 \frac{\mathbf{a_2}}{a_2} + x_3 \frac{\mathbf{a_3}}{a_3} \right).$$

This new function, $g(x_1, x_2, x_3)$, is now a function of three-variables, each of which has periodicity a_1, a_2, a_3 respectively: $g(x_1, x_2, x_3) = g(x_1 + a_1, x_2, x_3) = g(x_1, x_2 + a_2, x_3) = g(x_1, x_2, x_3 + a_3)$. If we write a series for g on the interval $[0, a_1]$ for x_1, we can define the following:

$$h^{\text{one}}(m_1, x_2, x_3) := \frac{1}{a_1} \int_0^{a_1} g(x_1, x_2, x_3) \cdot e^{-i2\pi \frac{m_1}{a_1} x_1} \, dx_1$$

And then we can write:

$$g(x_1, x_2, x_3) = \sum_{m_1=-\infty}^{\infty} h^{\text{one}}(m_1, x_2, x_3) \cdot e^{i2\pi \frac{m_1}{a_1} x_1}$$

Further defining:

$$h^{\text{two}}(m_1, m_2, x_3) := \frac{1}{a_2} \int_0^{a_2} h^{\text{one}}(m_1, x_2, x_3) \cdot e^{-i2\pi \frac{m_2}{a_2} x_2} \, dx_2$$

$$= \frac{1}{a_2} \int_0^{a_2} dx_2 \frac{1}{a_1} \int_0^{a_1} dx_1 g(x_1, x_2, x_3) \cdot e^{-i2\pi \left(\frac{m_1}{a_1} x_1 + \frac{m_2}{a_2} x_2 \right)}$$

We can write g once again as:

$$g(x_1, x_2, x_3) = \sum_{m_1=-\infty}^{\infty} \sum_{m_2=-\infty}^{\infty} h^{\text{two}}(m_1, m_2, x_3) \cdot e^{i2\pi \frac{m_1}{a_1} x_1} \cdot e^{i2\pi \frac{m_2}{a_2} x_2}$$

Finally applying the same for the third coordinate, we define:

$$h^{\text{three}}(m_1, m_2, m_3) := \frac{1}{a_3} \int_0^{a_3} h^{\text{two}}(m_1, m_2, x_3) \cdot e^{-i2\pi \frac{m_3}{a_3} x_3} \, dx_3$$

$$= \frac{1}{a_3} \int_0^{a_3} dx_3 \frac{1}{a_2} \int_0^{a_2} dx_2 \frac{1}{a_1} \int_0^{a_1} dx_1 g(x_1, x_2, x_3) \cdot e^{-i2\pi \left(\frac{m_1}{a_1} x_1 + \frac{m_2}{a_2} x_2 + \frac{m_3}{a_3} x_3 \right)}$$

We write g as:

$$g(x_1, x_2, x_3) = \sum_{m_1=-\infty}^{\infty} \sum_{m_2=-\infty}^{\infty} \sum_{m_3=-\infty}^{\infty} h^{\text{three}}(m_1, m_2, m_3) \cdot e^{i2\pi \frac{m_1}{a_1} x_1} \cdot e^{i2\pi \frac{m_2}{a_2} x_2} \cdot e^{i2\pi \frac{m_3}{a_3} x_3}$$

Re-arranging:

$$g(x_1, x_2, x_3) = \sum_{m_1, m_2, m_3 \in \mathbb{Z}} h^{\text{three}}(m_1, m_2, m_3) \cdot e^{i2\pi \left(\frac{m_1}{a_1} x_1 + \frac{m_2}{a_2} x_2 + \frac{m_3}{a_3} x_3 \right)}.$$

Now, every *reciprocal* lattice vector can be written as $\mathbf{K} = l_1 \mathbf{g}_1 + l_2 \mathbf{g}_2 + l_3 \mathbf{g}_3$, where li are integers and gi are the reciprocal lattice vectors, we can use the fact that $\mathbf{g_i} \cdot \mathbf{a_j} = 2\pi \delta_{ij}$ to calculate that for any arbitrary reciprocal lattice vector $\mathbf{K}$ and arbitrary vector in space $\mathbf{r}$, their scalar product is:

$$\mathbf{K} \cdot \mathbf{r} = (l_1 \mathbf{g}_1 + l_2 \mathbf{g}_2 + l_3 \mathbf{g}_3) \cdot \left(x_1 \frac{\mathbf{a}_1}{a_1} + x_2 \frac{\mathbf{a}_2}{a_2} + x_3 \frac{\mathbf{a}_3}{a_3} \right) = 2\pi \left(x_1 \frac{l_1}{a_1} + x_2 \frac{l_2}{a_2} + x_3 \frac{l_3}{a_3} \right).$$

And so it is clear that in our expansion, the sum is actually over reciprocal lattice vectors:

$$f(\mathbf{r}) = \sum_{\mathbf{K}} h(\mathbf{K}) \cdot e^{i\mathbf{K} \cdot \mathbf{r}},$$

where

$$h(\mathbf{K}) = \frac{1}{a_3} \int_0^{a_3} dx_3 \frac{1}{a_2} \int_0^{a_2} dx_2 \frac{1}{a_1} \int_0^{a_1} dx_1 f\left(x_1 \frac{\mathbf{a}_1}{a_1} + x_2 \frac{\mathbf{a}_2}{a_2} + x_3 \frac{\mathbf{a}_3}{a_3} \right) \cdot e^{-i\mathbf{K} \cdot \mathbf{r}}.$$

Assuming

$$\mathbf{r} = (x, y, z) = x_1 \frac{\mathbf{a}_1}{a_1} + x_2 \frac{\mathbf{a}_2}{a_2} + x_3 \frac{\mathbf{a}_3}{a_3},$$

we can solve this system of three linear equations for x, y, and z in terms of x_1, x_2 and x_3 in order to calculate the volume element in the original cartesian coordinate system. Once we have x, y, and z in terms of x_1, x_2 and x_3, we can calculate Jacobian determinant:

$$\begin{bmatrix} \dfrac{\partial x_1}{\partial x} & \dfrac{\partial x_1}{\partial y} & \dfrac{\partial x_1}{\partial z} \\[2mm] \dfrac{\partial x_2}{\partial x} & \dfrac{\partial x_2}{\partial y} & \dfrac{\partial x_2}{\partial z} \\[2mm] \dfrac{\partial x_3}{\partial x} & \dfrac{\partial x_3}{\partial y} & \dfrac{\partial x_3}{\partial z} \end{bmatrix}$$

which after some calculation and applying some non-trivial cross-product identities can be shown to be equal to:

$$\frac{a_1 a_2 a_3}{\mathbf{a_1} \cdot (\mathbf{a_2} \times \mathbf{a_3})}$$

(it may be advantageous for the sake of simplifying calculations, to work in such a cartesian coordinate system, in which it just so happens that $\mathbf{a}_1$ is parallel to the x axis, $\mathbf{a}_2$ lies in the x-y plane, and $\mathbf{a}_3$ has components of all three axes). The denominator is exactly the volume of the primitive unit cell which is enclosed by the three primitive-vectors $\mathbf{a}_1$, $\mathbf{a}_2$ and $\mathbf{a}_3$. In particular, we now know that

$$dx_1 \, dx_2 \, dx_3 = \frac{a_1 a_2 a_3}{\mathbf{a_1} \cdot (\mathbf{a_2} \times \mathbf{a_3})} \cdot dx \, dy \, dz.$$

We can write now $h(\mathbf{K})$ as an integral with the traditional coordinate system over the volume of the primitive cell, instead of with the x_1, x_2 and x_3 variables:

$$h(\mathbf{K}) = \frac{1}{\mathbf{a_1} \cdot (\mathbf{a_2} \times \mathbf{a_3})} \int_C d\mathbf{r} f(\mathbf{r}) \cdot e^{-i\mathbf{K}\cdot\mathbf{r}}$$

And C is the primitive unit cell, thus, $\mathbf{a_1} \cdot (\mathbf{a_2} \times \mathbf{a_3})$ is the volume of the primitive unit cell.

3.4.3   Hilbert space interpretation

Main article: Hilbert space

In the language of Hilbert spaces, the set of functions $\{\, e_n = e^{inx} \;;\; n \in \mathbf{Z} \}$ is an orthonormal basis for the space $L^2([-\pi, \pi])$ of square-integrable functions of $[-\pi, \pi]$. This space is actually a Hilbert space with an inner product given for any two elements f and g by

$$\langle f, g \rangle \overset{\text{def}}{=} \frac{1}{2\pi} \int_{-\pi}^{\pi} f(x)\overline{g(x)}\, dx.$$

The basic Fourier series result for Hilbert spaces can be written as

$$f = \sum_{n=-\infty}^{\infty} \langle f, e_n \rangle\, e_n.$$

This corresponds exactly to the complex exponential formulation given above. The version with sines and cosines is also

Sines and cosines form an orthonormal set, as illustrated above. The integral of sine, cosine and their product is zero (green and red areas are equal, and cancel out) when m, n or the functions are different, and pi only if m and n are equal, and the function used is the same.

justified with the Hilbert space interpretation. Indeed, the sines and cosines form an orthogonal set:

$$\int_{-\pi}^{\pi} \cos(mx)\,\cos(nx)\,dx = \pi\delta_{mn}, \quad m,n \geq 1,$$

$$\int_{-\pi}^{\pi} \sin(mx)\,\sin(nx)\,dx = \pi\delta_{mn}, \quad m,n \geq 1$$

(where δmn is the Kronecker delta), and

$$\int_{-\pi}^{\pi} \cos(mx)\,\sin(nx)\,dx = 0;$$

furthermore, the sines and cosines are orthogonal to the constant function **1**. An *orthonormal basis* for $L^2([-\pi,\pi])$ consisting of real functions is formed by the functions **1** and $\sqrt{2}\cos(nx)$, $\sqrt{2}\sin(nx)$ with $n = 1, 2,...$ The density of their span is a consequence of the Stone–Weierstrass theorem, but follows also from the properties of classical kernels like the Fejér kernel.

3.5   Properties

We say that f belongs to $C^k(\mathbb{T})$ if f is a 2π-periodic function on **R** which is k times differentiable, and its kth derivative is continuous.

- If f is a 2π-periodic odd function, then $an = 0$ for all n.

- If f is a 2π-periodic even function, then $bn = 0$ for all n.

- If f is integrable, $\lim_{|n|\to\infty} \hat{f}(n) = 0$, $\lim_{n\to+\infty} a_n = 0$ and $\lim_{n\to+\infty} b_n = 0$. This result is known as the Riemann–Lebesgue lemma.

- A doubly infinite sequence $\{an\}$ in $c_0(\mathbf{Z})$ is the sequence of Fourier coefficients of a function in $L^1([0, 2\pi])$ if and only if it is a convolution of two sequences in $\ell^2(\mathbf{Z})$. See [10]

- If $f \in C^1(\mathbb{T})$, then the Fourier coefficients $\widehat{f'}(n)$ of the derivative f' can be expressed in terms of the Fourier coefficients $\hat{f}(n)$ of the function f, via the formula $\widehat{f'}(n) = in\hat{f}(n)$.

- If $f \in C^k(\mathbb{T})$, then $\widehat{f^{(k)}}(n) = (in)^k \hat{f}(n)$. In particular, since $\widehat{f^{(k)}}(n)$ tends to zero, we have that $|n|^k \hat{f}(n)$ tends to zero, which means that the Fourier coefficients converge to zero faster than the kth power of n.

- Parseval's theorem. If f belongs to $L^2([-\pi, \pi])$, then $\sum_{n=-\infty}^{\infty} |\hat{f}(n)|^2 = \frac{1}{2\pi}\int_{-\pi}^{\pi} |f(x)|^2\,dx$.

- Plancherel's theorem. If $c_0, c_{\pm 1}, c_{\pm 2}, \dots$ are coefficients and $\sum_{n=-\infty}^{\infty} |c_n|^2 < \infty$ then there is a unique function $f \in L^2([-\pi, \pi])$ such that $\hat{f}(n) = c_n$ for every n.

- The first convolution theorem states that if f and g are in $L^1([-\pi, \pi])$, the Fourier series coefficients of the 2π-periodic convolution of f and g are given by:

$$[\widehat{f *_{2\pi} g}](n) = 2\pi \cdot \hat{f}(n) \cdot \hat{g}(n), \text{ [nb 4]}$$

where:

$$[f *_{2\pi} g](x) \overset{\text{def}}{=} \int_{-\pi}^{\pi} f(u) \cdot g[\text{pv}(x-u)] du, \quad \left(\text{ and } \underbrace{\text{pv}(x) \overset{\text{def}}{=} \text{Arg}\left(e^{ix}\right)}_{\text{value principal}} \right)$$

$$= \int_{-\pi}^{\pi} f(u) \cdot g(x-u) \, du, \qquad \text{2 is g(x) when } \pi\text{-periodic.}$$

$$= \int_{2\pi} f(u) \cdot g(x-u) \, du, \qquad \text{2 are functions both when } \pi\text{2 any over is integral the and -periodic, } \pi \text{interval.}$$

- The second convolution theorem states that the Fourier series coefficients of the product of f and g are given by the discrete convolution of the $\hat{f}$ and $\hat{g}$ sequences:

$$[\widehat{f \cdot g}](n) = [\hat{f} * \hat{g}](n).$$

3.5.1 Compact groups

Main articles: Compact group, Lie group and Peter–Weyl theorem

One of the interesting properties of the Fourier transform which we have mentioned, is that it carries convolutions to pointwise products. If that is the property which we seek to preserve, one can produce Fourier series on any compact group. Typical examples include those classical groups that are compact. This generalizes the Fourier transform to all spaces of the form $L^2(G)$, where G is a compact group, in such a way that the Fourier transform carries convolutions to pointwise products. The Fourier series exists and converges in similar ways to the $[-\pi, \pi]$ case.

An alternative extension to compact groups is the Peter–Weyl theorem, which proves results about representations of compact groups analogous to those about finite groups.

3.5.2 Riemannian manifolds

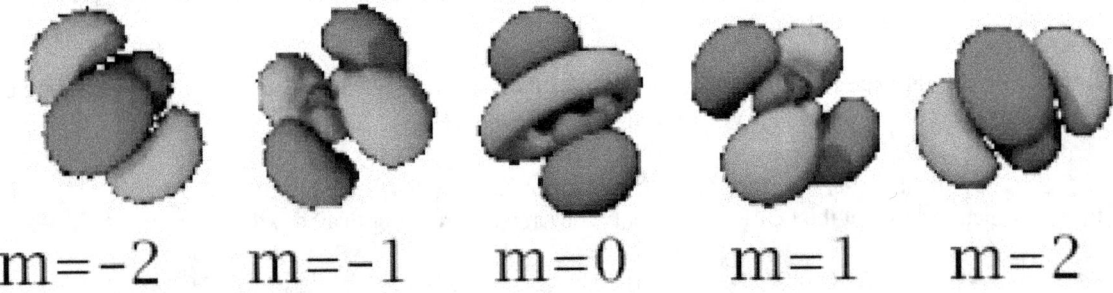

The atomic orbitals of chemistry are spherical harmonics and can be used to produce Fourier series on the sphere.

Main articles: Laplace operator and Riemannian manifold

If the domain is not a group, then there is no intrinsically defined convolution. However, if X is a compact Riemannian manifold, it has a Laplace–Beltrami operator. The Laplace–Beltrami operator is the differential operator that corresponds to Laplace operator for the Riemannian manifold X. Then, by analogy, one can consider heat equations on X. Since Fourier arrived at his basis by attempting to solve the heat equation, the natural generalization is to use the eigensolutions of the Laplace–Beltrami operator as a basis. This generalizes Fourier series to spaces of the type $L^2(X)$, where X is a Riemannian manifold. The Fourier series converges in ways similar to the $[-\pi, \pi]$ case. A typical example is to take X to be the sphere with the usual metric, in which case the Fourier basis consists of spherical harmonics.

3.5.3 Locally compact Abelian groups

Main article: Pontryagin duality

The generalization to compact groups discussed above does not generalize to noncompact, nonabelian groups. However, there is a straightfoward generalization to Locally Compact Abelian (LCA) groups.

This generalizes the Fourier transform to $L^1(G)$ or $L^2(G)$, where G is an LCA group. If G is compact, one also obtains a Fourier series, which converges similarly to the $[-\pi, \pi]$ case, but if G is noncompact, one obtains instead a Fourier integral. This generalization yields the usual Fourier transform when the underlying locally compact Abelian group is **R**.

3.6   Approximation and convergence of Fourier series

An important question for the theory as well as applications is that of convergence. In particular, it is often necessary in applications to replace the infinite series $\sum_{-\infty}^{\infty}$ by a finite one,

$$f_N(x) = \sum_{n=-N}^{N} \hat{f}(n) e^{inx}.$$

This is called a *partial sum*. We would like to know, in which sense does $fN(x)$ converge to $f(x)$ as $N \to \infty$.

3.6.1   Least squares property

We say that p is a trigonometric polynomial of degree N when it is of the form

$$p(x) = \sum_{n=-N}^{N} p_n e^{inx}.$$

Note that *fN is a trigonometric polynomial of degree* N. *Parseval's theorem implies that*

> **Theorem.** The trigonometric polynomial fN is the unique best trigonometric polynomial of degree N approximating $f(x)$, in the sense that, for any trigonometric polynomial $p \neq fN$ of degree N, we have
>
> $$\|f_N - f\|_2 < \|p - f\|_2,$$
>
> where the Hilbert space norm is defined as:
>
> $$\|g\|_2 = \sqrt{\frac{1}{2\pi} \int_{-\pi}^{\pi} |g(x)|^2 \, dx}.$$

3.6.2   Convergence

Main article: Convergence of Fourier series
See also: Gibbs phenomenon

Because of the least squares property, and because of the completeness of the Fourier basis, we obtain an elementary convergence result.

Theorem. If f belongs to $L^2([-\pi, \pi])$, then f∞ converges to f in $L^2([-\pi, \pi])$, that is, $\|f_N - f\|_2$ converges to 0 as $N \rightarrow \infty$.

We have already mentioned that if f is continuously differentiable, then $(i \cdot n)\hat{f}(n)$ is the nth Fourier coefficient of the derivative f'. It follows, essentially from the Cauchy–Schwarz inequality, that f∞ is absolutely summable. The sum of this series is a continuous function, equal to f, since the Fourier series converges in the mean to f:

Theorem. If $f \in C^1(\mathbb{T})$, then f∞ converges to f uniformly (and hence also pointwise.)

This result can be proven easily if f is further assumed to be C^2, since in that case $n^2 \hat{f}(n)$ tends to zero as $n \rightarrow \infty$. More generally, the Fourier series is absolutely summable, thus converges uniformly to f, provided that f satisfies a Hölder condition of order $\alpha > \frac{1}{2}$. In the absolutely summable case, the inequality $\sup_x |f(x) - f_N(x)| \le \sum_{|n|>N} |\hat{f}(n)|$ proves uniform convergence.

Many other results concerning the convergence of Fourier series are known, ranging from the moderately simple result that the series converges at x if f is differentiable at x, to Lennart Carleson's much more sophisticated result that the Fourier series of an L^2 function actually converges almost everywhere.

These theorems, and informal variations of them that don't specify the convergence conditions, are sometimes referred to generically as "Fourier's theorem" or "the Fourier theorem".[11][12][13][14]

3.6.3 Divergence

Since Fourier series have such good convergence properties, many are often surprised by some of the negative results. For example, the Fourier series of a continuous T-periodic function need not converge pointwise. The uniform boundedness principle yields a simple non-constructive proof of this fact.

In 1922, Andrey Kolmogorov published an article entitled "Une série de Fourier-Lebesgue divergente presque partout" in which he gave an example of a Lebesgue-integrable function whose Fourier series diverges almost everywhere. He later constructed an example of an integrable function whose Fourier series diverges everywhere (Katznelson 1976).

3.7   See also

- ATS theorem
- Dirichlet kernel
- Discrete Fourier transform
- Fast Fourier transform
- Fejér's theorem
- Fourier analysis
- Fourier sine and cosine series
- Fourier transform
- Gibbs phenomenon
- Laurent series – the substitution $q = e^{ix}$ transforms a Fourier series into a Laurent series, or conversely. This is used in the q-series expansion of the j-invariant.
- Multidimensional transform
- Spectral theory
- Sturm–Liouville theory

3.8   Notes

[1] These three did some important early work on the wave equation, especially D'Alembert. Euler's work in this area was mostly comtemporaneous/ in collaboration with Bernoulli, although the latter made some independent contributions to the theory of waves and vibrations (see here, pg.s 209 & 210,).

[2] Since the integral defining the Fourier transform of a periodic function is not convergent, it is necessary to view the periodic function and its transform as distributions. In this sense $\mathcal{F}\left\{e^{i\frac{2\pi nx}{P}}\right\}$ is a Dirac delta function, which is an example of a distribution.

[3] These words are not strictly Fourier's. Whilst the cited article does list the author as Fourier, a footnote indicates that the article was actually written by Poisson (that it was not written by Fourier is also clear from the consistent use of the third person to refer to him) and that it is, "for reasons of historical interest", presented as though it were Fourier's original memoire.

[4] The scale factor is always equal to the period, 2π in this case.

3.9   References

[1] Lejeune-Dirichlet, P. "Sur la convergence des séries trigonométriques qui servent à représenter une fonction arbitraire entre des limites données". (In French), transl. "On the convergence of trigonometric series which serve to represent an arbitrary function between two given limits". Journal für die reine und angewandte Mathematik, Vol. 4 (1829) pp. 157–169.

[2] "Ueber die Darstellbarkeit einer Function durch eine trigonometrische Reihe" [About the representability of a function by a trigonometric series]. *Habilitationsschrift, Göttingen; 1854. Abhandlungen der Königlichen Gesellschaft der Wissenschaften zu Göttingen, vol. 13, 1867.* Published posthumously for Riemann by Richard Dedekind *(in German). Archived from the original on 20 May 2008. Retrieved 19 May 2008.*

[3] D. Mascre, Bernhard Riemann: Posthumous Thesis on the Representation of Functions by Trigonometric Series (1867). Landmark Writings in Western Mathematics 1640–1940, Ivor Grattan-Guinness (ed.); pg. 492. Elsevier, 20 May 2005. Accessed 7 Dec 2012.</

[4] Theory of Complex Functions: Readings in Mathematics, by Reinhold Remmert; pg 29. Springer, 1991. Accessed 7 Dec 2012.

[5] Nerlove, Marc; Grether, David M.; Carvalho, Jose L. (1995). *Analysis of Economic Time Series. Economic Theory, Econometrics, and Mathematical Economics.* Elsevier. ISBN 0-12-515751-7.

[6] Flugge, Wilhelm (1957). *Statik und Dynamik der Schalen.* Berlin: Springer-Verlag.

[7] Dorf, Richard C.; Tallarida, Ronald J. (1993-07-15). *Pocket Book of Electrical Engineering Formulas* (1 ed.). Boca Raton,FL: CRC Press. pp. 171–174. ISBN 0849344735.

[8] Georgi P. Tolstov (1976). *Fourier Series.* Courier-Dover. ISBN 0-486-63317-9.

[9] "Gallica – Fourier, Jean-Baptiste-Joseph (1768–1830). Oeuvres de Fourier. 1888, pp. 218–219" (in French). Gallica.bnf.fr. 2007-10-15. Retrieved 2014-08-08.

[10] "fa.functional analysis - Characterizations of a linear subspace associated with Fourier series". MathOverflow. 2010-11-19. Retrieved 2014-08-08.

[11] William McC. Siebert (1985). *Circuits, signals, and systems.* MIT Press. p. 402. ISBN 978-0-262-19229-3.

[12] L. Marton and Claire Marton (1990). *Advances in Electronics and Electron Physics.* Academic Press. p. 369. ISBN 978-0-12-014650-5.

[13] Hans Kuzmany (1998). *Solid-state spectroscopy.* Springer. p. 14. ISBN 978-3-540-63913-8.

[14] Karl H. Pribram, Kunio Yasue, and Mari Jibu (1991). *Brain and perception.* Lawrence Erlbaum Associates. p. 26. ISBN 978-0-89859-995-4.

3.9.1   Further reading

- William E. Boyce and Richard C. DiPrima (2005). *Elementary Differential Equations and Boundary Value Problems* (8th ed.). New Jersey: John Wiley & Sons, Inc. ISBN 0-471-43338-1.

- Joseph Fourier, translated by Alexander Freeman (published 1822, translated 1878, re-released 2003). *The Analytical Theory of Heat*. Dover Publications. ISBN 0-486-49531-0. Check date values in: |date= (help) 2003 unabridged republication of the 1878 English translation by Alexander Freeman of Fourier's work *Théorie Analytique de la Chaleur*, originally published in 1822.

- Enrique A. Gonzalez-Velasco (1992). "Connections in Mathematical Analysis: The Case of Fourier Series". *American Mathematical Monthly* **99** (5): 427–441. doi:10.2307/2325087.

- Katznelson, Yitzhak (1976). "An introduction to harmonic analysis" (Second corrected ed.). New York: Dover Publications, Inc. ISBN 0-486-63331-4.

- Felix Klein, *Development of mathematics in the 19th century*. Mathsci Press Brookline, Mass, 1979. Translated by M. Ackerman from *Vorlesungen über die Entwicklung der Mathematik im 19 Jahrhundert*, Springer, Berlin, 1928.

- Walter Rudin (1976). *Principles of mathematical analysis* (3rd ed.). New York: McGraw-Hill, Inc. ISBN 0-07-054235-X.

- A. Zygmund (2002). *Trigonometric series* (third ed.). Cambridge: Cambridge University Press. ISBN 0-521-89053-5. The first edition was published in 1935.

3.10   External links

- thefouriertransform.com Fourier Series as a prelude to the Fourier Transform

- Characterizations of a linear subspace associated with Fourier series

- An interactive flash tutorial for the Fourier Series

- Phasor Phactory Allows custom control of the harmonic amplitudes for arbitrary terms

- Java applet shows Fourier series expansion of an arbitrary function

- Example problems – Examples of computing Fourier Series

- Hazewinkel, Michiel, ed. (2001), "Fourier series", *Encyclopedia of Mathematics*, Springer, ISBN 978-1-55608-010-4

- Weisstein, Eric W., "Fourier Series", *MathWorld*.

- Fourier Series Module by John H. Mathews

- Joseph Fourier – A site on Fourier's life which was used for the historical section of this article

- SFU.ca – 'Fourier Theorem'

Chapter 4

Generalized Fourier series

In mathematical analysis, many generalizations of Fourier series have proved to be useful. They are all special cases of decompositions over an orthonormal basis of an inner product space. Here we consider that of square-integrable functions defined on an interval of the real line, which is important, among others, for interpolation theory.

4.1   Definition

Consider a set of square-integrable functions with values in $\mathbb{F} = \mathbb{C}$ or $\mathbb{R}$,

$$\Phi = \{\varphi_n : [a,b] \to \mathbb{F}\}_{n=0}^{\infty},$$

which are pairwise orthogonal for the inner product

$$\langle f, g \rangle_w = \int_a^b f(x)\,\overline{g}(x)\,w(x)\,dx$$

where $w(x)$ is a weight function, and $\bar{\cdot}$ represents complex conjugation, i.e. $\overline{g}(x) = g(x)$ for $\mathbb{F} = \mathbb{R}$.

The **generalized Fourier series** of a square-integrable function $f \colon [a, b] \to \mathbb{F}$, with respect to Φ, is then

$$f(x) \sim \sum_{n=0}^{\infty} c_n \varphi_n(x),$$

where the coefficients are given by

$$c_n = \frac{\langle f, \varphi_n \rangle_w}{\|\varphi_n\|_w^2}.$$

If Φ is a complete set, i.e., an orthonormal basis of the space of all square-integrable functions on $[a, b]$, as opposed to a smaller orthonormal set, the relation $\sim$ becomes equality in the L^2 sense, more precisely modulo $|\cdot|w$ (not necessarily pointwise, nor almost everywhere).

4.2   Example (Fourier–Legendre series)

The Legendre polynomials are solutions to the Sturm–Liouville problem

$$\left((1 - x^2)P_n'(x)\right)' + n(n+1)P_n(x) = 0$$

and because of Sturm-Liouville theory, these polynomials are eigenfunctions of the problem and are solutions orthogonal with respect to the inner product above with unit weight. So we can form a generalized Fourier series (known as a Fourier–Legendre series) involving the Legendre polynomials, and

$$f(x) \sim \sum_{n=0}^{\infty} c_n P_n(x),$$

$$c_n = \frac{\langle f, P_n \rangle_w}{\|P_n\|_w^2}$$

As an example, let us calculate the Fourier–Legendre series for $f(x) = \cos x$ over $[-1, 1]$. Now,

$$c_0 = \sin 1 = \frac{\int_{-1}^{1} \cos x\, dx}{\int_{-1}^{1} (1)^2\, dx}$$

$$c_1 = 0 = \frac{\int_{-1}^{1} x \cos x\, dx}{\int_{-1}^{1} x^2\, dx} - \frac{0}{2/3}$$

$$c_2 = \frac{5}{2}(6 \cos 1 - 4 \sin 1) = \frac{\int_{-1}^{1} \frac{3x^2 - 1}{2} \cos x\, dx}{\int_{-1}^{1} \frac{9x^4 - 6x^2 + 1}{4}\, dx} = \frac{6 \cos 1 - 4 \sin 1}{2/5}$$

and a series involving these terms

$$c_2 P_2(x) + c_1 P_1(x) + c_0 P_0(x) = \frac{5}{2}(6 \cos 1 - 4 \sin 1)\left(\frac{3x^2 - 1}{2}\right) + \sin 1 (1)$$

$$= \left(\frac{45}{2} \cos 1 - 15 \sin 1\right) x^2 + 6 \sin 1 - \frac{15}{2} \cos 1$$

which differs from $\cos x$ by approximately 0.003, about 0. It may be advantageous to use such Fourier–Legendre series since the eigenfunctions are all polynomials and hence the integrals and thus the coefficients are easier to calculate.

4.3 Coefficient theorems

Some theorems on the coefficients cn include:

4.3.1 Bessel's inequality

$$\sum_{n=0}^{\infty} |c_n|^2 \leq \int_a^b |f(x)|^2\, dx.$$

4.3.2 Parseval's theorem

If Φ is a complete set,

$$\sum_{n=0}^{\infty} |c_n|^2 = \int_a^b |f(x)|^2\, dx.$$

4.4   See also

- Orthogonality
- Orthogonal function
- Eigenfunctions
- Vector space
- Function space
- Topological vector space
- Hilbert space
- Banach space

Chapter 5

Discrete-time Fourier transform

In mathematics, the **discrete-time Fourier transform** (**DTFT**) is a form of Fourier analysis that is applicable to the uniformly-spaced samples of a continuous function. The term *discrete-time* refers to the fact that the transform operates on discrete data (samples) whose interval often has units of time. From only the samples, it produces a function of frequency that is a periodic summation of the continuous Fourier transform of the original continuous function. Under certain theoretical conditions, described by the sampling theorem, the original continuous function can be recovered perfectly from the DTFT and thus from the original discrete samples. The DTFT itself is a continuous function of frequency, but discrete samples of it can be readily calculated via the discrete Fourier transform (DFT) (see Sampling the DTFT), which is by far the most common method of modern Fourier analysis.

Both transforms are invertible. The inverse DTFT is the original sampled data sequence. The inverse DFT is a periodic summation of the original sequence. The fast Fourier transform (FFT) is an algorithm for computing one cycle of the DFT, and its inverse produces one cycle of the inverse DFT.

5.1 Definition

The discrete-time Fourier transform of a discrete set of real or complex numbers: $x[n]$, for all integers n, is a Fourier series, which produces a periodic function of a frequency variable. When the frequency variable, ω, has normalized units of *radians/sample*, the periodicity is 2π, and the Fourier series is:

The utility of this frequency domain function is rooted in the Poisson summation formula. Let X(f) be the Fourier transform of any function, x(t), whose samples at some interval, T (*seconds*), are equal (or proportional to) the x[n] sequence, i.e. $T \cdot x(nT) = x[n]$. Then the periodic function represented by the Fourier series is a periodic summation of X(f). In terms of frequency f in hertz (*cycles/sec*):

The integer k has units of *cycles/sample*, and $1/T$ is the sample-rate, *fs* (*samples/sec*). So $X_1/T(f)$ comprises exact copies of $X(f)$ that are shifted by multiples of *fs* hertz and combined by addition. For sufficiently large *fs* the *k=0* term can be observed in the region [−*fs*/2, *fs*/2] with little or no distortion (aliasing) from the other terms. In Fig.1, the extremities of the distribution in the upper left corner are masked by aliasing in the periodic summation (lower left).

We also note that $e^{-i2\pi fTn}$ is the Fourier transform of $\delta(t-nT)$. Therefore, an alternative definition of DTFT is:[note 1]

The modulated Dirac comb function is a mathematical abstraction sometimes referred to as *impulse sampling*.[1]

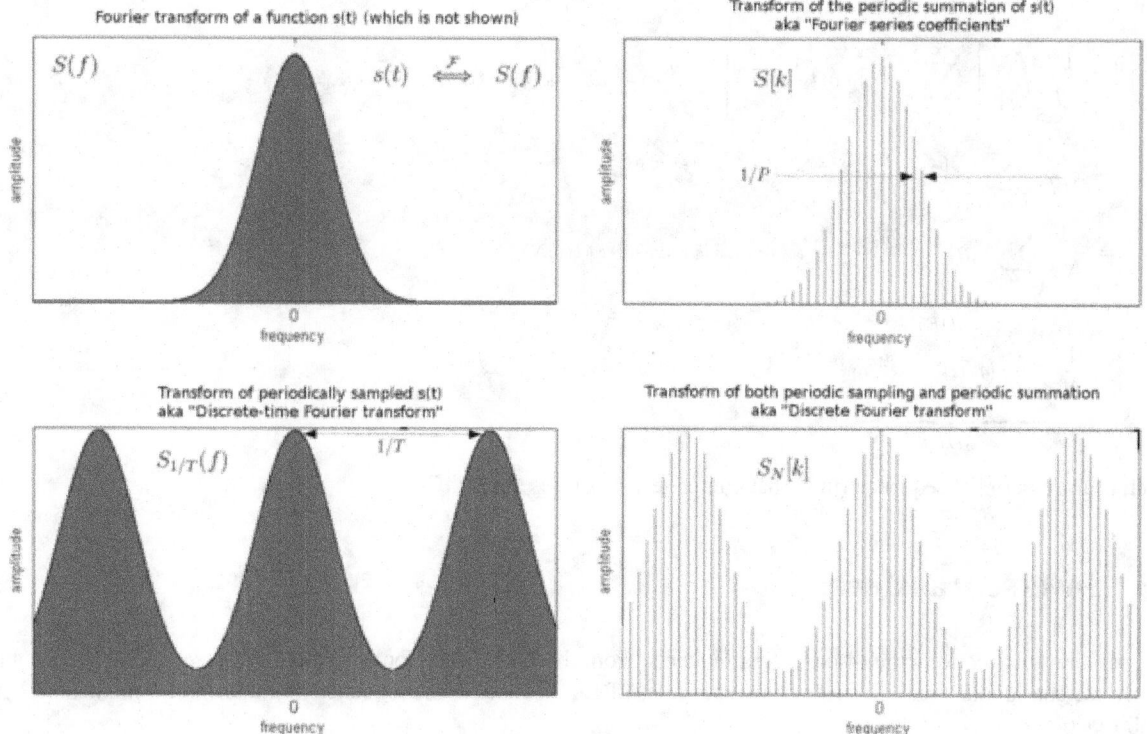

Fig 1. Depiction of a Fourier transform (upper left) and its periodic summation (DTFT) in the lower left corner. The lower right corner depicts samples of the DTFT that are computed by a discrete Fourier transform (DFT).

5.2   Periodic data

When the input data sequence $x[n]$ is N-periodic, **Eq.2** can be computationally reduced to a discrete Fourier transform (DFT), because:

- All the available information is contained within N samples.

- $X_{1/T}(f)$ converges to zero everywhere except integer multiples of $\frac{1}{NT}$, known as harmonic frequencies.

- The DTFT is periodic, so the maximum number of unique harmonic amplitudes is $\frac{1}{T} / \frac{1}{NT} = N$.

The kernel $x[n] \cdot e^{-i2\pi fTn}$ is N-periodic at the harmonic frequencies, $f = \frac{k}{NT}$. So $X_{1/T}(\frac{k}{NT})$ is an infinite summation of repetitious values, which does not converge for one or more values of k. But because of periodicity, we can reduce the limits of summation to any sequence of length N, without losing any information. The result is just a DFT. In order to interpret the DFT, it is helpful to expand the comb function, from **Eq.3**, which is now NT-periodic, into a Fourier series:

$$\underbrace{\sum_{n=-\infty}^{\infty} x[n] \cdot \delta(t - nT)}_{} = \underbrace{\sum_{k=-\infty}^{\infty} X[k] \cdot e^{i2\pi \frac{k}{NT} t}}_{\text{series Fourier}} \overset{\mathcal{F}}{\iff} \underbrace{\sum_{k=-\infty}^{\infty} X[k] \cdot \delta\left(f - \frac{k}{NT}\right)}_{\text{sequence periodic a of DTFT}},$$

which also shows that periodicity in the time domain causes the DTFT to become discontinuous and that it diverges at the harmonic frequencies. But the Fourier series coefficients that modulate the comb are finite, and the standard integral formula conveniently reduces to a DFT:

$$X[k] \stackrel{\text{def}}{=} \frac{1}{NT} \int_{NT} \left[\sum_{n=-\infty}^{\infty} x[n] \cdot \delta(t - nT) \right] e^{-i2\pi \frac{k}{NT} t}\, dt \qquad \text{(integral over any interval length } NT)$$

$$= \frac{1}{NT} \sum_{n=-\infty}^{\infty} x[n] \cdot \int_{NT} \delta(t - nT) \cdot e^{-i2\pi \frac{k}{NT} t}\, dt$$

$$= \frac{1}{NT} \underbrace{\sum_{N} x[n] \cdot e^{-i2\pi \frac{k}{N} n}}_{DFT} \qquad \text{(sum over any } n \text{ length of -sequence } N)$$

$$= \frac{1}{N} \underbrace{\sum_{N} x(nT) \cdot e^{-i2\pi \frac{k}{N} n}}_{DFT},$$

which is an N-periodic sequence (in k) that completely describes the DTFT.

5.3 Inverse transform

An operation that recovers the discrete data sequence from the DTFT function is called an *inverse DTFT*. For instance, the inverse continuous Fourier transform of both sides of **Eq.3** produces the sequence in the form of a modulated Dirac comb function:

$$\sum_{n=-\infty}^{\infty} x[n] \cdot \delta(t - nT) = \mathcal{F}^{-1}\left\{ X_{1/T}(f) \right\} \stackrel{\text{def}}{=} \int_{-\infty}^{\infty} X_{1/T}(f) \cdot e^{i2\pi ft}\, df.$$

However, noting that $X_1/T(f)$ is periodic, all the necessary information is contained within any interval of length $1/T$. In both **Eq.1** and **Eq.2**, the summations over n are a Fourier series, with coefficients x[n]. The standard formulas for the Fourier coefficients are also the inverse transforms:

$$x[n] = T \int_{\frac{1}{T}} X_{1/T}(f) \cdot e^{i2\pi fnT}\, df \qquad \text{(integral over any interval length } 1/T)$$

$$= \frac{1}{2\pi} \int_{2\pi} X_{2\pi}(\omega) \cdot e^{i\omega n}\, d\omega \qquad \text{(integral over any interval length } 2\pi)$$

5.4 Sampling the DTFT

When the DTFT is continuous, a common practice is to compute an arbitrary number of samples (N) of one cycle of the periodic function $X1/T$:

$$\underbrace{X_{1/T}\left(\frac{k}{NT}\right)}_{X_k} = \sum_{n=-\infty}^{\infty} x[n] \cdot e^{-i2\pi \frac{kn}{N}} \qquad k = 0, \ldots, N-1$$

$$= \underbrace{\sum_{N} x_N[n] \cdot e^{-i2\pi \frac{kn}{N}}}_{DFT}, \qquad \text{(sum over any } n \text{ length of -sequence } N)$$

where xN is a periodic summation:

$$x_N[n] \stackrel{\text{def}}{=} \sum_{m=-\infty}^{\infty} x[n - mN].$$

The xN sequence is the inverse DFT. Thus, our sampling of the DTFT causes the inverse transform to become periodic.

In order to evaluate one cycle of xN numerically, we require a finite-length x[n] sequence. For instance, a long sequence might be truncated by a window function of length L resulting in two cases worthy of special mention: $L \leq N$ and $L = I \cdot N$, for some integer I (typically 6 or 8). For notational simplicity, consider the $x[n]$ values below to represent the modified values.

When $L = I \cdot N$ a cycle of xN reduces to a summation of I *blocks* of length N. This goes by various names, such as "multi-block windowing" and "window presum-DFT". [2] [3] A good way to understand/motivate the technique is to recall that decimation of sampled data in one domain (time or frequency) produces aliasing in the other, and vice versa. The xN summation is mathematically equivalent to aliasing, leading to decimation in frequency, leaving only DTFT samples least affected by spectral leakage. That is usually a priority when implementing an FFT filter-bank (channelizer). With a conventional window function of length L, scalloping loss would be unacceptable. So multi-block windows are created using FIR filter design tools. Their frequency profile is flat at the highest point and falls off quickly at the midpoint between the remaining DTFT samples. The larger the value of parameter I the better the potential performance. We note that the same results can be obtained by computing and decimating an L-length DFT, but that is not computationally efficient.

When $L \leq N$ the DFT is usually written in this more familiar form:

$$X_k = \sum_{n=0}^{N-1} x[n] \cdot e^{-i2\pi \frac{kn}{N}}.$$

In order to take advantage of a fast Fourier transform algorithm for computing the DFT, the summation is usually performed over all N terms, even though N-L of them are zeros. Therefore, the case $L < N$ is often referred to as "zero-padding".

Spectral leakage, which increases as L decreases, is detrimental to certain important performance metrics, such as resolution of multiple frequency components and the amount of noise measured by each DTFT sample. But those things don't always matter, for instance when the x[n] sequence is a noiseless sinusoid (or a constant), shaped by a window function. Then it is a common practice to use *zero-padding* to graphically display and compare the detailed leakage patterns of window functions. To illustrate that for a rectangular window, consider the sequence:

$$x[n] = e^{i2\pi \frac{1}{8}n}, \quad \text{and } L = 64.$$

The two figures below are plots of the magnitude of two different sized DFTs, as indicated in their labels. In both cases, the dominant component is at the signal frequency: $f = 1/8 = 0.125$. Also visible on the right is the spectral leakage pattern of the $L = 64$ rectangular window. The illusion on the left is a result of sampling the DTFT at all of its zero-crossings. Rather than the DTFT of a finite-length sequence, it gives the impression of an infinitely long sinusoidal sequence. Contributing factors to the illusion are the use of a rectangular window, and the choice of a frequency (1/8 = 8/64) with exactly 8 (an integer) cycles per 64 samples.

5.5 Convolution

The convolution theorem for sequences is:

$$x * y = \text{DTFT}^{-1}\left[\text{DTFT}\{x\} \cdot \text{DTFT}\{y\}\right].$$

An important special case is the circular convolution of sequences x and y defined by $xN * y$ where xN is a periodic summation. The discrete-frequency nature of DTFT$\{xN\}$ "selects" only discrete values from the continuous function DTFT$\{y\}$, which results in considerable simplification of the inverse transform. As shown at Convolution theorem#Functions of a discrete variable... sequences:

$$x_N * y = \text{DTFT}^{-1}\left[\text{DTFT}\{x_N\} \cdot \text{DTFT}\{y\}\right] = \text{DFT}^{-1}\left[\text{DFT}\{x_N\} \cdot \text{DFT}\{y_N\}\right].$$

For x and y sequences whose non-zero duration is less than or equal to N, a final simplification is:

$$x_N * y = \text{DFT}^{-1}\left[\text{DFT}\{x\} \cdot \text{DFT}\{y\}\right].$$

The significance of this result is expounded at circular convolution and Fast convolution algorithms.

5.6 Relationship to the Z-transform

The bilateral Z-transform is defined by:

$$X(z) = \sum_{n=-\infty}^{\infty} x[n]\, z^{-n}, \text{ where z is a complex variable.}$$

On the unit circle, z is constrained to values of the form $e^{i\omega}$. Then one cycle of $X(e^{i\omega})$, $0 \leq \omega \leq 2\pi$ is equivalent to one period of the DTFT. What varies with sample-rate is the width of a signal's spectral distribution. When the width exceeds 2π, because of a sub-Nyquist rate, the distribution fills the circle, and aliasing occurs. With a DTFT in units of hertz (**Eq.2**), it's not the bandwidth that changes, but the periodicity of the aliases.

5.6.1 Alternative notation

The notation, $X(e^{i\omega})$, is also often used to denote a normalized DTFT (**Eq.1**), which has several desirable features:

1. highlights the periodicity property, and

2. helps distinguish between the DTFT and the underlying Fourier transform of x(t); that is, X(f) (or X(ω)), and

3. emphasizes the relationship of the DTFT to the Z-transform.

However, its relevance is obscured when the DTFT is expressed as its equivalent periodic summation. So the notation X(ω) is also commonly used, as in the table below.

5.7 Table of discrete-time Fourier transforms

Some common transform pairs are shown in the table below. The following notation applies:

- $\omega = 2\pi f T$ is a real number representing continuous angular frequency (in radians per sample). (f is in cycles/sec, and T is in sec/sample.) In all cases in the table, the DTFT is 2π-periodic (in ω).

- $X_{2\pi}(\omega)$ designates a function defined on $-\infty < \omega < \infty$.

- X(ω) designates a function defined on $-\pi < \omega \leq \pi$, and zero elsewhere. Then:

$$X_{2\pi}(\omega) \stackrel{\text{def}}{=} \sum_{k=-\infty}^{\infty} X(\omega - 2\pi k).$$

- $\delta(\omega)$ is the Dirac delta function
- $\mathrm{sinc}(t)$ is the normalized sinc function
- $\mathrm{rect}(t)$ is the rectangle function
- $\mathrm{tri}(t)$ is the triangle function
- n is an integer representing the discrete-time domain (in samples)
- $u[n]$ is the discrete-time unit step function
- $\delta[n]$ is the Kronecker delta $\delta_{n,\,0}$

5.8 Properties

This table shows some mathematical operations in the time domain and the corresponding effects in the frequency domain.

- $*$ is the discrete convolution of two sequences
- $x[n]^*$ is the complex conjugate of $x[n]$
- $X(e^{i\omega})$ is the alternative notation (described above) for $X(\omega)$

5.9 See also

- Multidimensional transform
- Zak transform

5.10 Notes

[1] In fact **Eq.2** is often justified as follows:

$$\mathcal{F}\left\{\sum_{n=-\infty}^{\infty} T \cdot x(nT) \cdot \delta(t - nT)\right\} = \mathcal{F}\left\{x(t) \cdot T \sum_{n=-\infty}^{\infty} \delta(t - nT)\right\}$$
$$= X(f) * \mathcal{F}\left\{T \sum_{n=-\infty}^{\infty} \delta(t - nT)\right\}$$
$$= X(f) * \sum_{k=-\infty}^{\infty} \delta\left(f - \frac{k}{T}\right)$$
$$= \sum_{k=-\infty}^{\infty} X\left(f - \frac{k}{T}\right).$$

5.11 Citations

[1] Rao, R. *Signals and Systems*. Prentice-Hall Of India Pvt. Limited. ISBN 9788120338593.

[2] Gumas, Charles Constantine (July 1997). "Window-presum FFT achieves high-dynamic range, resolution". *Personal Engineering & Instrumentation News*: 58–64.

[3] Lyons, Richard G. (June 2008). "DSP Tricks: Building a practical spectrum analyzer". EE Times.

5.12 References

- Alan V. Oppenheim and Ronald W. Schafer (1999). *Discrete-Time Signal Processing* (2nd ed.). Prentice Hall Signal Processing Series. ISBN 0-13-754920-2.

- William McC. Siebert (1986). *Circuits, Signals, and Systems*. MIT Electrical Engineering and Computer Science Series. Cambridge, MA: MIT Press.

- Boaz Porat. *A Course in Digital Signal Processing*. John Wiley and Sons. pp. 27–29 and 104–105. ISBN 0-471-14961-6.

Chapter 6

Discrete Fourier transform

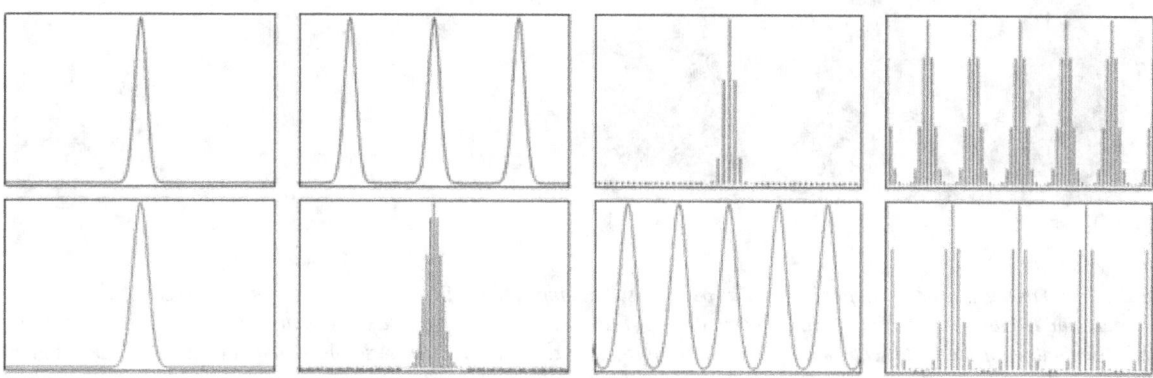

Relationship between the (continuous) Fourier transform and the discrete Fourier transform. Left column: A continuous function (top) and its Fourier transform (bottom). Center-left column: Periodic summation of the original function (top). Fourier transform (bottom) is zero except at discrete points. The inverse transform is a sum of sinusoids called Fourier series. Center-right column: Original function is discretized (multiplied by a Dirac comb) (top). Its Fourier transform (bottom) is a periodic summation (DTFT) of the original transform. Right column: The DFT (bottom) computes discrete samples of the continuous DTFT. The inverse DFT (top) is a periodic summation of the original samples. The FFT algorithm computes one cycle of the DFT and its inverse is one cycle of the DFT inverse.

In mathematics, the **discrete Fourier transform (DFT)** converts a finite list of equally spaced samples of a function into the list of coefficients of a finite combination of complex sinusoids, ordered by their frequencies, that has those same sample values. It can be said to convert the sampled function from its original domain (often time or position along a line) to the frequency domain.

The input samples are complex numbers (in practice, usually real numbers), and the output coefficients are complex as well. The frequencies of the output sinusoids are integer multiples of a fundamental frequency, whose corresponding period is the length of the sampling interval. The combination of sinusoids obtained through the DFT is therefore periodic with that same period. The DFT differs from the discrete-time Fourier transform (DTFT) in that its input **and** output sequences are both finite; it is therefore said to be the Fourier analysis of finite-domain (or periodic) discrete-time functions.

The DFT is the most important discrete transform, used to perform Fourier analysis in many practical applications.[1] In digital signal processing, the function is any quantity or signal that varies over time, such as the pressure of a sound wave, a radio signal, or daily temperature readings, sampled over a finite time interval (often defined by a window function[2]). In image processing, the samples can be the values of pixels along a row or column of a raster image. The DFT is also used to efficiently solve partial differential equations, and to perform other operations such as convolutions or multiplying large integers.

Since it deals with a finite amount of data, it can be implemented in computers by numerical algorithms or even dedicated hardware. These implementations usually employ efficient fast Fourier transform (FFT) algorithms;[3] so much so that the terms "FFT" and "DFT" are often used interchangeably. Prior to its current usage, the "FFT" initialism may have also

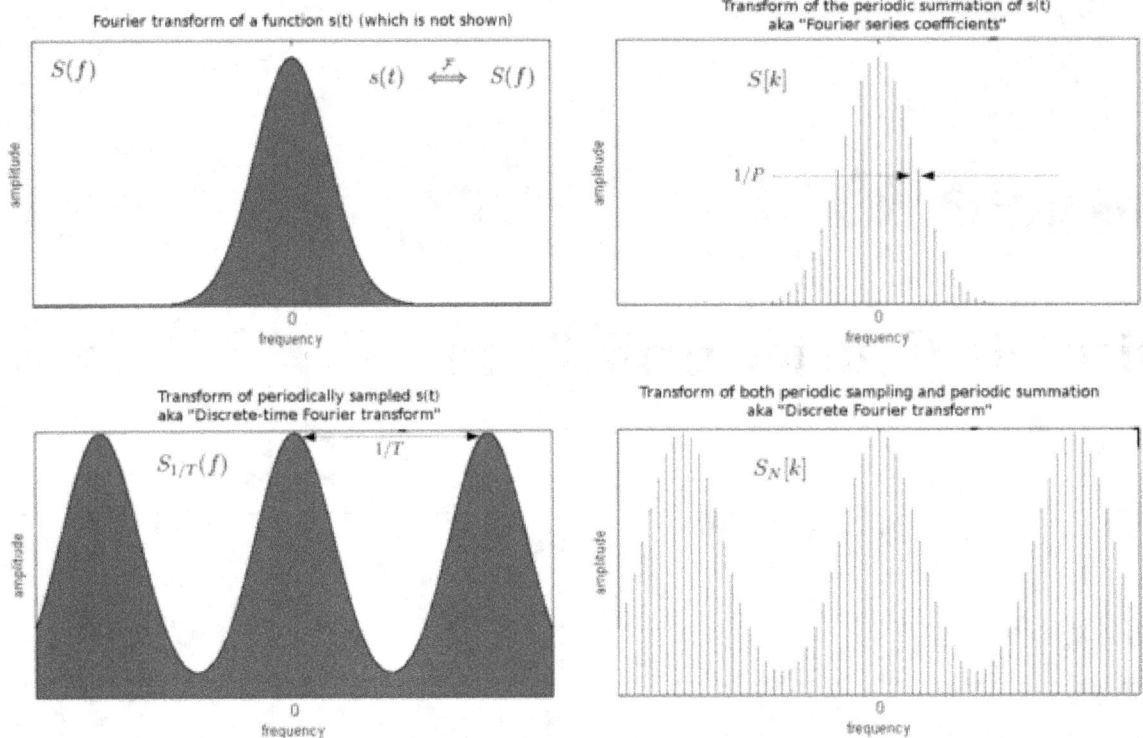

*Depiction of a Fourier transform (upper left) and its periodic summation (DTFT) in the lower left corner. The spectral sequences at (a) upper right and (b) lower right are respectively computed from (a) one cycle of the periodic summation of s(t) and (b) one cycle of the periodic summation of the s(nT) sequence. The respective formulas are (a) the Fourier series integral and (b) the **DFT** summation. Its similarities to the original transform, S(f), and its relative computational ease are often the motivation for computing a DFT sequence.*

been used for the ambiguous term "finite Fourier transform".

6.1   Definition

The sequence of N complex numbers $x_0, x_1, \ldots, x_{N-1}$ is transformed into an N-periodic sequence of complex numbers:

Because of periodicity, the customary domain of **k** actually computed is $[0, N-1]$. That is always the case when the DFT is implemented via the Fast Fourier transform algorithm. But other common domains are $[-N/2, N/2-1]$ (N even) and $[-(N-1)/2, (N-1)/2]$ (N odd), as when the left and right halves of an FFT output sequence are swapped.[4]

The transform is sometimes denoted by the symbol $\mathcal{F}$, as in $\mathbf{X} = \mathcal{F}\{\mathbf{x}\}$ or $\mathcal{F}(\mathbf{x})$ or $\mathcal{F}\mathbf{x}$.[note 2]

Eq.1 can be interpreted or derived in various ways, for example:

- It completely describes the discrete-time Fourier transform (DTFT) of an N-periodic sequence, which comprises only discrete frequency components. (Using the DTFT with periodic data)

- It can also provide uniformly spaced samples of the continuous DTFT of a finite length sequence. (Sampling the DTFT)

- It is the cross correlation of the *input* sequence, *xn*, and a complex sinusoid at frequency *k/N*. Thus it acts like a matched filter for that frequency.

- It is the discrete analogy of the formula for the coefficients of a Fourier series:

which is also **N**-periodic. In the domain $n \in [0, N-1]$, this is the **inverse transform** of **Eq.1**. In this interpretation, each X_k is a complex number that encodes both amplitude and phase of a sinusoidal component $\left(e^{j2\pi kn/N}\right)$ of function x_n . The sinusoid's frequency is k cycles per N samples. Its amplitude and phase are:

$$|X_k|/N = \sqrt{\mathrm{Re}(X_k)^2 + \mathrm{Im}(X_k)^2}/N$$

$$\arg(X_k) = \mathrm{atan2}\left(\mathrm{Im}(X_k), \mathrm{Re}(X_k)\right) = -i \ln\left(\frac{X_k}{|X_k|}\right),$$

where atan2 is the two-argument form of the arctan function.

The normalization factor multiplying the DFT and IDFT (here 1 and $1/N$) and the signs of the exponents are merely conventions, and differ in some treatments. The only requirements of these conventions are that the DFT and IDFT have opposite-sign exponents and that the product of their normalization factors be $1/N$. A normalization of $\sqrt{1/N}$ for both the DFT and IDFT, for instance, makes the transforms unitary.

In the following discussion the terms "sequence" and "vector" will be considered interchangeable.

Using Euler's Formula, it can be derived further to the forms commonly used in Engineering and Computer Science.

Fourier Transform:

Inverse Fourier Transform:

N = number of time samples we have

n = current sample we're considering (0...N-1)

$\mathbf{x_n}$ = value of the signal at time n

k = current frequency we're considering (0 Hertz up to N-1 Hertz)

$\mathbf{X_k}$ = amount of frequency k in the signal (Amplitude and Phase, a complex number)

6.2   Properties

6.2.1   Completeness

The discrete Fourier transform is an invertible, linear transformation

$$\mathcal{F}\colon \mathbb{C}^N \to \mathbb{C}^N$$

with $\mathbb{C}$ denoting the set of complex numbers. In other words, for any $N > 0$, an N-dimensional complex vector has a DFT and an IDFT which are in turn N-dimensional complex vectors.

6.2.2 Orthogonality

The vectors $u_k = \left[e^{\frac{2\pi i}{N} kn} \mid n = 0, 1, \ldots, N-1 \right]^T$ form an orthogonal basis over the set of N-dimensional complex vectors:

$$u_k^T u_{k'}^* = \sum_{n=0}^{N-1} \left(e^{\frac{2\pi i}{N} kn} \right) \left(e^{\frac{2\pi i}{N} (-k')n} \right) = \sum_{n=0}^{N-1} e^{\frac{2\pi i}{N} (k-k')n} = N \, \delta_{kk'}$$

where $\delta_{kk'}$ is the Kronecker delta. (In the last step, the summation is trivial if $k = k'$, where it is $1+1+\cdots=N$, and otherwise is a geometric series that can be explicitly summed to obtain zero.) This orthogonality condition can be used to derive the formula for the IDFT from the definition of the DFT, and is equivalent to the unitarity property below.

6.2.3 The Plancherel theorem and Parseval's theorem

If Xk and Yk are the DFTs of xn and yn respectively then the Parseval's theorem states:

$$\sum_{n=0}^{N-1} x_n y_n^* = \frac{1}{N} \sum_{k=0}^{N-1} X_k Y_k^*$$

where the star denotes complex conjugation. Plancherel theorem is a special case of the Parseval's theorem and states:

$$\sum_{n=0}^{N-1} |x_n|^2 = \frac{1}{N} \sum_{k=0}^{N-1} |X_k|^2.$$

These theorems are also equivalent to the unitary condition below.

6.2.4 Periodicity

The periodicity can be shown directly from the definition:

$$X_{k+N} \overset{\text{def}}{=} \sum_{n=0}^{N-1} x_n e^{-\frac{2\pi i}{N}(k+N)n} = \sum_{n=0}^{N-1} x_n e^{-\frac{2\pi i}{N} kn} \underbrace{e^{-2\pi i n}}_{1} = \sum_{n=0}^{N-1} x_n e^{-\frac{2\pi i}{N} kn} = X_k.$$

Similarly, it can be shown that the IDFT formula leads to a periodic extension.

6.2.5 Shift theorem

Multiplying x_n by a *linear phase* $e^{\frac{2\pi i}{N} nm}$ for some integer m corresponds to a *circular shift* of the output X_k : X_k is replaced by X_{k-m}, where the subscript is interpreted modulo N (i.e., periodically). Similarly, a circular shift of the input x_n corresponds to multiplying the output X_k by a linear phase. Mathematically, if $\{x_n\}$ represents the vector **x** then

$$\mathcal{F}(\{x_n\})_k = X_k$$

$$\mathcal{F}(\{x_n \cdot e^{\frac{2\pi i}{N} nm}\})_k = X_{k-m}$$

$$\mathcal{F}(\{x_{n-m}\})_k = X_k \cdot e^{-\frac{2\pi i}{N} km}$$

6.2.6 Circular convolution theorem and cross-correlation theorem

The convolution theorem for the discrete-time Fourier transform indicates that a convolution of two infinite sequences can be obtained as the inverse transform of the product of the individual transforms. An important simplification occurs when the sequences are of finite length, **N**. In terms of the DFT and inverse DFT, it can be written as follows:

$$\mathcal{F}^{-1}\left\{\mathbf{X} \cdot \mathbf{Y}\right\}_n = \sum_{l=0}^{N-1} x_l \cdot (y_N)_{n-l} \overset{\text{def}}{=} (\mathbf{x} * \mathbf{y_N})_n ,$$

which is the convolution of the **x** sequence with a **y** sequence extended by periodic summation:

$$(\mathbf{y_N})_n \overset{\text{def}}{=} \sum_{p=-\infty}^{\infty} y_{(n-pN)} = y_{n(mod N)}.$$

Similarly, the cross-correlation of **x** and **y$_N$** is given by:

$$\mathcal{F}^{-1}\left\{\mathbf{X}^* \cdot \mathbf{Y}\right\}_n = \sum_{l=0}^{N-1} x_l^* \cdot (y_N)_{n+l} \overset{\text{def}}{=} (\mathbf{x} \star \mathbf{y_N})_n .$$

When either sequence contains a string of zeros, of length L, L+1 of the circular convolution outputs are equivalent to values of **x** * **y**. Methods have also been developed to use this property as part of an efficient process that constructs **x** * **y** with an **x** or **y** sequence potentially much longer than the practical transform size (**N**). Two such methods are called overlap-save and overlap-add.[5] The efficiency results from the fact that a direct evaluation of either summation (above) requires $O(N^2)$ operations for an output sequence of length N. An indirect method, using transforms, can take advantage of the $O(N \log N)$ efficiency of the fast Fourier transform (FFT) to achieve much better performance. Furthermore, convolutions can be used to efficiently compute DFTs via Rader's FFT algorithm and Bluestein's FFT algorithm.

6.2.7 Convolution theorem duality

It can also be shown that:

$$\mathcal{F}\left\{\mathbf{x} \cdot \mathbf{y}\right\}_k \overset{\text{def}}{=} \sum_{n=0}^{N-1} x_n \cdot y_n \cdot e^{-\frac{2\pi i}{N}kn}$$

$$= \tfrac{1}{N}(\mathbf{X} * \mathbf{Y_N})_k, \text{ which is the circular convolution of } \mathbf{X} \text{ and } \mathbf{Y} .$$

6.2.8 Trigonometric interpolation polynomial

The trigonometric interpolation polynomial

$$p(t) = \tfrac{1}{N}\left[X_0 + X_1 e^{2\pi it} + \cdots + X_{N/2-1}e^{(N/2-1)2\pi it} + X_{N/2}\cos(N\pi t) + X_{N/2+1}e^{(-N/2+1)2\pi it} \cdots + X_{N-1}e^{-2\pi it}\right]$$
for N even ,

$$p(t) = \tfrac{1}{N}\left[X_0 + X_1 e^{2\pi it} + \cdots + X_{\lfloor N/2 \rfloor}e^{\lfloor N/2 \rfloor 2\pi it} + X_{\lfloor N/2 \rfloor+1}e^{-\lfloor N/2 \rfloor 2\pi it} + \cdots + X_{N-1}e^{-2\pi it}\right] \text{ for}$$
N odd,

where the coefficients Xk are given by the DFT of xn above, satisfies the interpolation property $p(n/N) = x_n$ for $n = 0, \ldots, N-1$.

For even N, notice that the Nyquist component $\frac{X_{N/2}}{N}\cos(N\pi t)$ is handled specially.

This interpolation is *not unique*: aliasing implies that one could add N to any of the complex-sinusoid frequencies (e.g. changing e^{-it} to $e^{i(N-1)t}$) without changing the interpolation property, but giving *different* values in between the x_n points. The choice above, however, is typical because it has two useful properties. First, it consists of sinusoids whose frequencies have the smallest possible magnitudes: the interpolation is bandlimited. Second, if the x_n are real numbers, then $p(t)$ is real as well.

In contrast, the most obvious trigonometric interpolation polynomial is the one in which the frequencies range from 0 to $N - 1$ (instead of roughly $-N/2$ to $+N/2$ as above), similar to the inverse DFT formula. This interpolation does *not* minimize the slope, and is *not* generally real-valued for real x_n ; its use is a common mistake.

6.2.9 The unitary DFT

Another way of looking at the DFT is to note that in the above discussion, the DFT can be expressed as a Vandermonde matrix, introduced by Sylvester in 1867,

$$
\mathbf{F} = \begin{bmatrix} \omega_N^{0 \cdot 0} & \omega_N^{0 \cdot 1} & \cdots & \omega_N^{0 \cdot (N-1)} \\ \omega_N^{1 \cdot 0} & \omega_N^{1 \cdot 1} & \cdots & \omega_N^{1 \cdot (N-1)} \\ \vdots & \vdots & \ddots & \vdots \\ \omega_N^{(N-1) \cdot 0} & \omega_N^{(N-1) \cdot 1} & \cdots & \omega_N^{(N-1) \cdot (N-1)} \end{bmatrix}
$$

where

$$
\omega_N = e^{-2\pi i / N}
$$

is a primitive Nth root of unity.

The inverse transform is then given by the inverse of the above matrix,

$$
\mathbf{F}^{-1} = \frac{1}{N} \mathbf{F}^*
$$

With unitary normalization constants $1/\sqrt{N}$, the DFT becomes a unitary transformation, defined by a unitary matrix:

$$
\mathbf{U} = \mathbf{F}/\sqrt{N}
$$

$$
\mathbf{U}^{-1} = \mathbf{U}^*
$$

$$
|\det(\mathbf{U})| = 1
$$

where *det()* is the determinant function. The determinant is the product of the eigenvalues, which are always ± 1 or $\pm i$ as described below. In a real vector space, a unitary transformation can be thought of as simply a rigid rotation of the coordinate system, and all of the properties of a rigid rotation can be found in the unitary DFT.

The orthogonality of the DFT is now expressed as an orthonormality condition (which arises in many areas of mathematics as described in root of unity):

$$
\sum_{m=0}^{N-1} U_{km} U_{mn}^* = \delta_{kn}
$$

If $\mathbf{X}$ is defined as the unitary DFT of the vector $\mathbf{x}$, then

$$X_k = \sum_{n=0}^{N-1} U_{kn} x_n$$

and the Plancherel theorem is expressed as

$$\sum_{n=0}^{N-1} x_n y_n^* = \sum_{k=0}^{N-1} X_k Y_k^*$$

If we view the DFT as just a coordinate transformation which simply specifies the components of a vector in a new coordinate system, then the above is just the statement that the dot product of two vectors is preserved under a unitary DFT transformation. For the special case $\mathbf{x} = \mathbf{y}$, this implies that the length of a vector is preserved as well—this is just Parseval's theorem,

$$\sum_{n=0}^{N-1} |x_n|^2 = \sum_{k=0}^{N-1} |X_k|^2$$

A consequence of the circular convolution theorem is that the DFT matrix F diagonalizes any circulant matrix.

6.2.10   Expressing the inverse DFT in terms of the DFT

A useful property of the DFT is that the inverse DFT can be easily expressed in terms of the (forward) DFT, via several well-known "tricks". (For example, in computations, it is often convenient to only implement a fast Fourier transform corresponding to one transform direction and then to get the other transform direction from the first.)

First, we can compute the inverse DFT by reversing the inputs (Duhamel *et al.*, 1988):

$$\mathcal{F}^{-1}(\{x_n\}) = \mathcal{F}(\{x_{N-n}\})/N$$

(As usual, the subscripts are interpreted modulo N; thus, for $n = 0$, we have $x_{N-0} = x_0$.)

Second, one can also conjugate the inputs and outputs:

$$\mathcal{F}^{-1}(\mathbf{x}) = \mathcal{F}(\mathbf{x}^*)^*/N$$

Third, a variant of this conjugation trick, which is sometimes preferable because it requires no modification of the data values, involves swapping real and imaginary parts (which can be done on a computer simply by modifying pointers). Define swap(x_n) as x_n with its real and imaginary parts swapped—that is, if $x_n = a + bi$ then swap(x_n) is $b + ai$. Equivalently, swap(x_n) equals ix_n^*. Then

$$\mathcal{F}^{-1}(\mathbf{x}) = \mathrm{swap}(\mathcal{F}(\mathrm{swap}(\mathbf{x})))/N$$

That is, the inverse transform is the same as the forward transform with the real and imaginary parts swapped for both input and output, up to a normalization (Duhamel *et al.*, 1988).

The conjugation trick can also be used to define a new transform, closely related to the DFT, that is involutory—that is, which is its own inverse. In particular, $T(\mathbf{x}) = \mathcal{F}(\mathbf{x}^*)/\sqrt{N}$ is clearly its own inverse: $T(T(\mathbf{x})) = \mathbf{x}$. A closely related involutory transformation (by a factor of $(1+i)/\sqrt{2}$) is $H(\mathbf{x}) = \mathcal{F}((1+i)\mathbf{x}^*)/\sqrt{2N}$, since the $(1+i)$ factors in $H(H(\mathbf{x}))$ cancel the 2. For real inputs $\mathbf{x}$, the real part of $H(\mathbf{x})$ is none other than the discrete Hartley transform, which is also involutory.

6.2.11   Eigenvalues and eigenvectors

The eigenvalues of the DFT matrix are simple and well-known, whereas the eigenvectors are complicated, not unique, and are the subject of ongoing research.

Consider the unitary form **U** defined above for the DFT of length N, where

$$\mathbf{U}_{m,n} = \frac{1}{\sqrt{N}}\omega_N^{(m-1)(n-1)} = \frac{1}{\sqrt{N}}e^{-\frac{2\pi i}{N}(m-1)(n-1)}.$$

This matrix satisfies the matrix polynomial equation:

$$\mathbf{U}^4 = \mathbf{I}.$$

This can be seen from the inverse properties above: operating **U** twice gives the original data in reverse order, so operating **U** four times gives back the original data and is thus the identity matrix. This means that the eigenvalues λ satisfy the equation:

$$\lambda^4 = 1.$$

Therefore, the eigenvalues of **U** are the fourth roots of unity: λ is $+1$, -1, $+i$, or $-i$.

Since there are only four distinct eigenvalues for this $N \times N$ matrix, they have some multiplicity. The multiplicity gives the number of linearly independent eigenvectors corresponding to each eigenvalue. (Note that there are N independent eigenvectors; a unitary matrix is never defective.)

The problem of their multiplicity was solved by McClellan and Parks (1972), although it was later shown to have been equivalent to a problem solved by Gauss (Dickinson and Steiglitz, 1982). The multiplicity depends on the value of N modulo 4, and is given by the following table:

Otherwise stated, the characteristic polynomial of **U** is:

$$\det(\lambda I - \mathbf{U}) = (\lambda - 1)^{\left\lfloor \frac{N+4}{4} \right\rfloor}(\lambda + 1)^{\left\lfloor \frac{N+2}{4} \right\rfloor}(\lambda + i)^{\left\lfloor \frac{N+1}{4} \right\rfloor}(\lambda - i)^{\left\lfloor \frac{N-1}{4} \right\rfloor}.$$

No simple analytical formula for general eigenvectors is known. Moreover, the eigenvectors are not unique because any linear combination of eigenvectors for the same eigenvalue is also an eigenvector for that eigenvalue. Various researchers have proposed different choices of eigenvectors, selected to satisfy useful properties like orthogonality and to have "simple" forms (e.g., McClellan and Parks, 1972; Dickinson and Steiglitz, 1982; Grünbaum, 1982; Atakishiyev and Wolf, 1997; Candan *et al.*, 2000; Hanna *et al.*, 2004; Gurevich and Hadani, 2008).

A straightforward approach is to discretize an eigenfunction of the continuous Fourier transform, of which the most famous is the Gaussian function. Since periodic summation of the function means discretizing its frequency spectrum and discretization means periodic summation of the spectrum, the discretized and periodically summed Gaussian function yields an eigenvector of the discrete transform:

- $F(m) = \sum_{k \in \mathbb{Z}} \exp\left(-\frac{\pi \cdot (m+N \cdot k)^2}{N}\right).$

A closed form expression for the series is not known, but it converges rapidly.

Two other simple closed-form analytical eigenvectors for special DFT period N were found (Kong, 2008):

For DFT period $N = 2L + 1 = 4K + 1$, where K is an integer, the following is an eigenvector of DFT:

- $F(m) = \prod_{s=K+1}^{L} \left[\cos\left(\frac{2\pi}{N}m\right) - \cos\left(\frac{2\pi}{N}s\right)\right]$

For DFT period $N = 2L = 4K$, where K is an integer, the following is an eigenvector of DFT:

- $F(m) = \sin\left(\frac{2\pi}{N}m\right)\prod_{s=K+1}^{L-1} \left[\cos\left(\frac{2\pi}{N}m\right) - \cos\left(\frac{2\pi}{N}s\right)\right]$

The choice of eigenvectors of the DFT matrix has become important in recent years in order to define a discrete analogue of the fractional Fourier transform—the DFT matrix can be taken to fractional powers by exponentiating the eigenvalues (e.g., Rubio and Santhanam, 2005). For the continuous Fourier transform, the natural orthogonal eigenfunctions are the Hermite functions, so various discrete analogues of these have been employed as the eigenvectors of the DFT, such as the Kravchuk polynomials (Atakishiyev and Wolf, 1997). The "best" choice of eigenvectors to define a fractional discrete Fourier transform remains an open question, however.

6.2.12   Uncertainty principle

If the random variable Xk is constrained by

$$\sum_{n=0}^{N-1} |X_n|^2 = 1\,,$$

then

$$P_n = |X_n|^2$$

may be considered to represent a discrete probability mass function of n, with an associated probability mass function constructed from the transformed variable,

$$Q_m = N|x_m|^2\,.$$

For the case of continuous functions $P(x)$ and $Q(k)$, the Heisenberg uncertainty principle states that

$$D_0(X)D_0(x) \geq \frac{1}{16\pi^2}$$

where $D_0(X)$ and $D_0(x)$ are the variances of $|X|^2$ and $|x|^2$ respectively, with the equality attained in the case of a suitably normalized Gaussian distribution. Although the variances may be analogously defined for the DFT, an analogous uncertainty principle is not useful, because the uncertainty will not be shift-invariant. Still, a meaningful uncertainty principle has been introduced by Massar and Spindel.[6]

However, the Hirschman entropic uncertainty will have a useful analog for the case of the DFT.[7] The Hirschman uncertainty principle is expressed in terms of the Shannon entropy of the two probability functions.

In the discrete case, the Shannon entropies are defined as

$$H(X) = -\sum_{n=0}^{N-1} P_n \ln P_n$$

and

$$H(x) = -\sum_{m=0}^{N-1} Q_m \ln Q_m \,,$$

and the entropic uncertainty principle becomes[7]

$$H(X) + H(x) \geq \ln(N)\,.$$

The equality is obtained for P_n equal to translations and modulations of a suitably normalized Kronecker comb of period A where A is any exact integer divisor of N. The probability mass function Q_m will then be proportional to a suitably translated Kronecker comb of period $B = N/A$.[7]

There is also a well-known deterministic uncertainty principle that uses signal sparsity (or the number of non-zero coefficients).[8] Let $\|x\|_0$ and $\|X\|_0$ be the number of non-zero elements of the time and frequency sequences $x_0, x_1, \ldots, x_{N-1}$ and $X_0, X_1, \ldots, X_{N-1}$, respectively. Then,

$$N \leq \|x\|_0 \cdot \|X\|_0.$$

As an immediate consequence of the inequality of arithmetic and geometric means, one also has $2\sqrt{N} \leq \|x\|_0 + \|X\|_0$. Both uncertainty principles were shown to be tight for specifically-chosen "picket-fence" sequences (discrete impulse trains), and find practical use for signal recovery applications.[8]

6.2.13   The real-input DFT

If $x_0, \ldots, x_{N-1}$ are real numbers, as they often are in practical applications, then the DFT obeys the symmetry:

$$X_{N-k} \equiv X_{-k} = X_k^*, \text{ where } X^* \text{ denotes complex conjugation.}$$

It follows that *X0* and *XN/2* are real-valued, and the remainder of the DFT is completely specified by just *N/2-1* complex numbers.

6.3   Generalized DFT (shifted and non-linear phase)

It is possible to shift the transform sampling in time and/or frequency domain by some real shifts a and b, respectively. This is sometimes known as a **generalized DFT** (or **GDFT**), also called the **shifted DFT** or **offset DFT**, and has analogous properties to the ordinary DFT:

$$X_k = \sum_{n=0}^{N-1} x_n e^{-\frac{2\pi i}{N}(k+b)(n+a)} \qquad k = 0, \ldots, N-1.$$

Most often, shifts of $1/2$ (half a sample) are used. While the ordinary DFT corresponds to a periodic signal in both time and frequency domains, $a = 1/2$ produces a signal that is anti-periodic in frequency domain ($X_{k+N} = -X_k$) and vice versa for $b = 1/2$. Thus, the specific case of $a = b = 1/2$ is known as an *odd-time odd-frequency* discrete Fourier transform (or O^2 DFT). Such shifted transforms are most often used for symmetric data, to represent different boundary symmetries, and for real-symmetric data they correspond to different forms of the discrete cosine and sine transforms.

Another interesting choice is $a = b = -(N-1)/2$, which is called the **centered DFT** (or **CDFT**). The centered DFT has the useful property that, when N is a multiple of four, all four of its eigenvalues (see above) have equal multiplicities (Rubio and Santhanam, 2005)[9]

The term GDFT is also used for the non-linear phase extensions of DFT. Hence, GDFT method provides a generalization for constant amplitude orthogonal block transforms including linear and non-linear phase types. GDFT is a framework to improve time and frequency domain properties of the traditional DFT, e.g. auto/cross-correlations, by the addition of the properly designed phase shaping function (non-linear, in general) to the original linear phase functions (Akansu and Agirman-Tosun, 2010).[10]

The discrete Fourier transform can be viewed as a special case of the z-transform, evaluated on the unit circle in the complex plane; more general z-transforms correspond to *complex* shifts a and b above.

6.4   Multidimensional DFT

The ordinary DFT transforms a one-dimensional sequence or array x_n that is a function of exactly one discrete variable n. The multidimensional DFT of a multidimensional array $x_{n_1,n_2,\ldots,n_d}$ that is a function of d discrete variables $n_\ell = 0, 1, \ldots, N_\ell - 1$ for ℓ in $1, 2, \ldots, d$ is defined by:

$$X_{k_1,k_2,\ldots,k_d} = \sum_{n_1=0}^{N_1-1} \left(\omega_{N_1}^{k_1 n_1} \sum_{n_2=0}^{N_2-1} \left(\omega_{N_2}^{k_2 n_2} \cdots \sum_{n_d=0}^{N_d-1} \omega_{N_d}^{k_d n_d} \cdot x_{n_1,n_2,\ldots,n_d} \right) \right),$$

where $\omega_{N_\ell} = \exp(-2\pi i / N_\ell)$ as above and the d output indices run from $k_\ell = 0, 1, \ldots, N_\ell - 1$. This is more compactly expressed in vector notation, where we define $\mathbf{n} = (n_1, n_2, \ldots, n_d)$ and $\mathbf{k} = (k_1, k_2, \ldots, k_d)$ as d-dimensional vectors of indices from 0 to $\mathbf{N} - 1$, which we define as $\mathbf{N} - 1 = (N_1 - 1, N_2 - 1, \ldots, N_d - 1)$:

$$X_{\mathbf{k}} = \sum_{\mathbf{n}=0}^{\mathbf{N}-1} e^{-2\pi i \mathbf{k} \cdot (\mathbf{n}/\mathbf{N})} x_{\mathbf{n}},$$

where the division $\mathbf{n}/\mathbf{N}$ is defined as $\mathbf{n}/\mathbf{N} = (n_1/N_1, \ldots, n_d/N_d)$ to be performed element-wise, and the sum denotes the set of nested summations above.

The inverse of the multi-dimensional DFT is, analogous to the one-dimensional case, given by:

$$x_{\mathbf{n}} = \frac{1}{\prod_{\ell=1}^{d} N_\ell} \sum_{\mathbf{k}=0}^{\mathbf{N}-1} e^{2\pi i \mathbf{n} \cdot (\mathbf{k}/\mathbf{N})} X_{\mathbf{k}}.$$

As the one-dimensional DFT expresses the input x_n as a superposition of sinusoids, the multidimensional DFT expresses the input as a superposition of plane waves, or multidimensional sinusoids. The direction of oscillation in space is $\mathbf{k}/\mathbf{N}$. The amplitudes are $X_{\mathbf{k}}$. This decomposition is of great importance for everything from digital image processing (two-dimensional) to solving partial differential equations. The solution is broken up into plane waves.

The multidimensional DFT can be computed by the composition of a sequence of one-dimensional DFTs along each dimension. In the two-dimensional case x_{n_1,n_2} the N_1 independent DFTs of the rows (i.e., along n_2) are computed first to form a new array y_{n_1,k_2}. Then the N_2 independent DFTs of y along the columns (along n_1) are computed to form the final result X_{k_1,k_2}. Alternatively the columns can be computed first and then the rows. The order is immaterial because the nested summations above commute.

An algorithm to compute a one-dimensional DFT is thus sufficient to efficiently compute a multidimensional DFT. This approach is known as the *row-column* algorithm. There are also intrinsically multidimensional FFT algorithms.

6.4.1   The real-input multidimensional DFT

For input data $x_{n_1,n_2,\ldots,n_d}$ consisting of real numbers, the DFT outputs have a conjugate symmetry similar to the one-dimensional case above:

$$X_{k_1,k_2,\ldots,k_d} = X^*_{N_1-k_1,N_2-k_2,\ldots,N_d-k_d},$$

where the star again denotes complex conjugation and the ℓ-th subscript is again interpreted modulo N_ℓ (for $\ell = 1, 2, \ldots, d$).

6.5 Applications

The DFT has seen wide usage across a large number of fields; we only sketch a few examples below (see also the references at the end). All applications of the DFT depend crucially on the availability of a fast algorithm to compute discrete Fourier transforms and their inverses, a fast Fourier transform.

6.5.1 Spectral analysis

When the DFT is used for signal spectral analysis, the $\{x_n\}$ sequence usually represents a finite set of uniformly spaced time-samples of some signal $x(t)$, where t represents time. The conversion from continuous time to samples (discrete-time) changes the underlying Fourier transform of x(t) into a discrete-time Fourier transform (DTFT), which generally entails a type of distortion called aliasing. Choice of an appropriate sample-rate (see *Nyquist rate*) is the key to minimizing that distortion. Similarly, the conversion from a very long (or infinite) sequence to a manageable size entails a type of distortion called *leakage*, which is manifested as a loss of detail (a.k.a. resolution) in the DTFT. Choice of an appropriate sub-sequence length is the primary key to minimizing that effect. When the available data (and time to process it) is more than the amount needed to attain the desired frequency resolution, a standard technique is to perform multiple DFTs, for example to create a spectrogram. If the desired result is a power spectrum and noise or randomness is present in the data, averaging the magnitude components of the multiple DFTs is a useful procedure to reduce the variance of the spectrum (also called a periodogram in this context); two examples of such techniques are the Welch method and the Bartlett method; the general subject of estimating the power spectrum of a noisy signal is called spectral estimation.

A final source of distortion (or perhaps *illusion*) is the DFT itself, because it is just a discrete sampling of the DTFT, which is a function of a continuous frequency domain. That can be mitigated by increasing the resolution of the DFT. That procedure is illustrated at Sampling the DTFT.

- The procedure is sometimes referred to as *zero-padding*, which is a particular implementation used in conjunction with the fast Fourier transform (FFT) algorithm. The inefficiency of performing multiplications and additions with zero-valued "samples" is more than offset by the inherent efficiency of the FFT.

- As already noted, leakage imposes a limit on the inherent resolution of the DTFT. So there is a practical limit to the benefit that can be obtained from a fine-grained DFT.

6.5.2 Filter bank

See FFT filter banks and Sampling the DTFT.

6.5.3 Data compression

The field of digital signal processing relies heavily on operations in the frequency domain (i.e. on the Fourier transform). For example, several lossy image and sound compression methods employ the discrete Fourier transform: the signal is cut into short segments, each is transformed, and then the Fourier coefficients of high frequencies, which are assumed to be unnoticeable, are discarded. The decompressor computes the inverse transform based on this reduced number of Fourier coefficients. (Compression applications often use a specialized form of the DFT, the discrete cosine transform or sometimes the modified discrete cosine transform.) Some relatively recent compression algorithms, however, use wavelet transforms, which give a more uniform compromise between time and frequency domain than obtained by chopping data into segments and transforming each segment. In the case of JPEG2000, this avoids the spurious image features that appear when images are highly compressed with the original JPEG.

6.5.4   Partial differential equations

Discrete Fourier transforms are often used to solve partial differential equations, where again the DFT is used as an approximation for the Fourier series (which is recovered in the limit of infinite N). The advantage of this approach is that it expands the signal in complex exponentials e^{inx}, which are eigenfunctions of differentiation: $d/dx \, e^{inx} = in \, e^{inx}$. Thus, in the Fourier representation, differentiation is simple—we just multiply by $i \, n$. (Note, however, that the choice of n is not unique due to aliasing; for the method to be convergent, a choice similar to that in the trigonometric interpolation section above should be used.) A linear differential equation with constant coefficients is transformed into an easily solvable algebraic equation. One then uses the inverse DFT to transform the result back into the ordinary spatial representation. Such an approach is called a spectral method.

6.5.5   Polynomial multiplication

Suppose we wish to compute the polynomial product $c(x) = a(x) \cdot b(x)$. The ordinary product expression for the coefficients of c involves a linear (acyclic) convolution, where indices do not "wrap around." This can be rewritten as a cyclic convolution by taking the coefficient vectors for $a(x)$ and $b(x)$ with constant term first, then appending zeros so that the resultant coefficient vectors $\mathbf{a}$ and $\mathbf{b}$ have dimension $d > \deg(a(x)) + \deg(b(x))$. Then,

$$\mathbf{c} = \mathbf{a} * \mathbf{b}$$

Where $\mathbf{c}$ is the vector of coefficients for $c(x)$, and the convolution operator $*$ is defined so

$$c_n = \sum_{m=0}^{d-1} a_m b_{n-m \bmod d} \qquad\qquad n = 0, 1 \ldots, d-1$$

But convolution becomes multiplication under the DFT:

$$\mathcal{F}(\mathbf{c}) = \mathcal{F}(\mathbf{a})\mathcal{F}(\mathbf{b})$$

Here the vector product is taken elementwise. Thus the coefficients of the product polynomial $c(x)$ are just the terms 0, ..., $\deg(a(x)) + \deg(b(x))$ of the coefficient vector

$$\mathbf{c} = \mathcal{F}^{-1}(\mathcal{F}(\mathbf{a})\mathcal{F}(\mathbf{b})).$$

With a fast Fourier transform, the resulting algorithm takes O ($N \log N$) arithmetic operations. Due to its simplicity and speed, the Cooley–Tukey FFT algorithm, which is limited to composite sizes, is often chosen for the transform operation. In this case, d should be chosen as the smallest integer greater than the sum of the input polynomial degrees that is factorizable into small prime factors (e.g. 2, 3, and 5, depending upon the FFT implementation).

Multiplication of large integers

The fastest known algorithms for the multiplication of very large integers use the polynomial multiplication method outlined above. Integers can be treated as the value of a polynomial evaluated specifically at the number base, with the coefficients of the polynomial corresponding to the digits in that base. After polynomial multiplication, a relatively low-complexity carry-propagation step completes the multiplication.

Convolution

When data is convolved with a function with wide support, such as for downsampling by a large sampling ratio, because of the Convolution theorem and the FFT algorithm, it may be faster to transform it, multiply pointwise by the transform of the filter and then reverse transform it. Alternatively, a good filter is obtained by simply truncating the transformed data and re-transforming the shortened data set.

6.6 Some discrete Fourier transform pairs

6.7 Generalizations

6.7.1 Representation theory

For more details on this topic, see Representation theory of finite groups § Discrete Fourier transform.

The DFT can be interpreted as the complex-valued representation theory of the finite cyclic group. In other words, a sequence of n complex numbers can be thought of as an element of n-dimensional complex space $\mathbf{C}^n$ or equivalently a function f from the finite cyclic group of order n to the complex numbers, $\mathbf{Z}n \to \mathbf{C}$. So f is a class function on the finite cyclic group, and thus can be expressed as a linear combination of the irreducible characters of this group, which are the roots of unity.

From this point of view, one may generalize the DFT to representation theory generally, or more narrowly to the representation theory of finite groups.

More narrowly still, one may generalize the DFT by either changing the target (taking values in a field other than the complex numbers), or the domain (a group other than a finite cyclic group), as detailed in the sequel.

6.7.2 Other fields

Main articles: Discrete Fourier transform (general) and Number-theoretic transform

Many of the properties of the DFT only depend on the fact that $e^{-\frac{2\pi i}{N}}$ is a primitive root of unity, sometimes denoted ω_N or W_N (so that $\omega_N^N = 1$). Such properties include the completeness, orthogonality, Plancherel/Parseval, periodicity, shift, convolution, and unitarity properties above, as well as many FFT algorithms. For this reason, the discrete Fourier transform can be defined by using roots of unity in fields other than the complex numbers, and such generalizations are commonly called *number-theoretic transforms* (NTTs) in the case of finite fields. For more information, see number-theoretic transform and discrete Fourier transform (general).

6.7.3 Other finite groups

Main article: Fourier transform on finite groups

The standard DFT acts on a sequence $x_0, x_1, \ldots, xN_{-1}$ of complex numbers, which can be viewed as a function $\{0, 1, \ldots, N - 1\} \to \mathbf{C}$. The multidimensional DFT acts on multidimensional sequences, which can be viewed as functions

$$\{0, 1, \ldots, N_1 - 1\} \times \cdots \times \{0, 1, \ldots, N_d - 1\} \to \mathbb{C}.$$

This suggests the generalization to Fourier transforms on arbitrary finite groups, which act on functions $G \to \mathbf{C}$ where G is a finite group. In this framework, the standard DFT is seen as the Fourier transform on a cyclic group, while the multidimensional DFT is a Fourier transform on a direct sum of cyclic groups.

6.8   Alternatives

Main article: Discrete wavelet transform
For more details on this topic, see Discrete wavelet transform § Comparison with Fourier transform.

There are various alternatives to the DFT for various applications, prominent among which are wavelets. The analog of the DFT is the discrete wavelet transform (DWT). From the point of view of time–frequency analysis, a key limitation of the Fourier transform is that it does not include *location* information, only *frequency* information, and thus has difficulty in representing transients. As wavelets have location as well as frequency, they are better able to represent location, at the expense of greater difficulty representing frequency. For details, see comparison of the discrete wavelet transform with the discrete Fourier transform.

6.9   See also

- Companion matrix

- DFT matrix

- Fast Fourier transform

- FFTPACK

- FFTW

- Generalizations of Pauli matrices

- List of Fourier-related transforms

- Multidimensional transform

- Zak transform

6.10   Notes

[1] In this context, it is common to define ω to be the Nth primitive root of unity, $\omega = e^{-2\pi i/N}$, to obtain the following form:

$$X_k = \sum_{n=0}^{N-1} x_n \cdot \omega^{kn}$$

[2] As a linear transformation on a finite-dimensional vector space, the DFT expression can also be written in terms of a DFT matrix; when scaled appropriately it becomes a unitary matrix and the Xk can thus be viewed as coefficients of x in an orthonormal basis.

6.11   Citations

[1] Strang, Gilbert (May–June 1994). "Wavelets". *American Scientist* **82** (3): 253. Retrieved 8 October 2013. This is the most important numerical algorithm of our lifetime...

[2] Sahidullah, Md.; Saha, Goutam (Feb 2013). "A Novel Windowing Technique for Efficient Computation of MFCC for Speaker Recognition".*IEEE Signal Processing Letters***20**(2): 149–152. arXiv:1206.2437. Bibcode:2013ISPL...20..149S.doi:10.1109/L

[3] Cooley et al., 1969

[4] "Shift zero-frequency component to center of spectrum - MATLAB fftshift". *http://www.mathworks.com/*. Natick, MA 01760: The MathWorks, Inc. Retrieved 10 March 2014.

[5] T. G. Stockham, Jr., "High-speed convolution and correlation," in 1966 *Proc. AFIPS Spring Joint Computing Conf.* Reprinted in Digital Signal Processing, L. R. Rabiner and C. M. Rader, editors, New York: IEEE Press, 1972.

[6] Massar, S.; Spindel, P. (2008). "Uncertainty Relation for the Discrete Fourier Transform". *Physical Review Letters* **100** (19). arXiv:0710.0723. Bibcode:2008PhRvL.100s0401M. doi:10.1103/PhysRevLett.100.190401.

[7] DeBrunner, Victor; Havlicek, Joseph P.; Przebinda, Tomasz; Özaydin, Murad (2005). "Entropy-Based Uncertainty Measures for $L^2(\mathbb{R}^n)$, $\ell^2(\mathbb{Z})$, and $\ell^2(\mathbb{Z}/N\mathbb{Z})$ With a Hirschman Optimal Transform for $\ell^2(\mathbb{Z}/N\mathbb{Z})$ " (PDF). *IEEE Transactions on Signal Processing* **53** (8): 2690. Bibcode:2005ITSP...53.2690D. doi:10.1109/TSP.2005.850329. Retrieved 2011-06-23.

[8] Donoho, D.L.; Stark, P.B (1989). "Uncertainty principles and signal recovery". *SIAM Journal on Applied Mathematics* **49** (3): 906–931. doi:10.1137/0149053.

[9] Santhanam, Balu; Santhanam, Thalanayar S. "*Discrete Gauss-Hermite functions and eigenvectors of the centered discrete Fourier transform*", Proceedings of the 32nd IEEE *International Conference on Acoustics, Speech, and Signal Processing* (ICASSP 2007, SPTM-P12.4), vol. III, pp. 1385-1388.

[10] Akansu, Ali N.; Agirman-Tosun, Handan "*Generalized Discrete Fourier Transform With Nonlinear Phase*", IEEE *Transactions on Signal Processing*, vol. 58, no. 9, pp. 4547-4556, Sept. 2010.

6.12   References

- Brigham, E. Oran (1988). *The fast Fourier transform and its applications*. Englewood Cliffs, N.J.: Prentice Hall. ISBN 0-13-307505-2.

- Oppenheim, Alan V.; Schafer, R. W.; and Buck, J. R. (1999). *Discrete-time signal processing*. Upper Saddle River, N.J.: Prentice Hall. ISBN 0-13-754920-2.

- Smith, Steven W. (1999). "Chapter 8: The Discrete Fourier Transform". *The Scientist and Engineer's Guide to Digital Signal Processing* (Second ed.). San Diego, Calif.: California Technical Publishing. ISBN 0-9660176-3-3.

- Cormen, Thomas H.; Charles E. Leiserson; Ronald L. Rivest; Clifford Stein (2001). "Chapter 30: Polynomials and the FFT". *Introduction to Algorithms* (Second ed.). MIT Press and McGraw-Hill. pp. 822–848. ISBN 0-262-03293-7. esp. section 30.2: The DFT and FFT, pp. 830–838.

- P. Duhamel, B. Piron, and J. M. Etcheto (1988). "On computing the inverse DFT". *IEEE Trans. Acoust., Speech and Sig. Processing* **36** (2): 285–286. doi:10.1109/29.1519.

- J. H. McClellan and T. W. Parks (1972). "Eigenvalues and eigenvectors of the discrete Fourier transformation". *IEEE Trans. Audio Electroacoust.* **20** (1): 66–74. doi:10.1109/TAU.1972.1162342.

- Bradley W. Dickinson and Kenneth Steiglitz (1982). "Eigenvectors and functions of the discrete Fourier transform". *IEEE Trans. Acoust., Speech and Sig. Processing* **30** (1): 25–31. doi:10.1109/TASSP.1982.1163843. (Note that this paper has an apparent typo in its table of the eigenvalue multiplicities: the $+i/-i$ columns are interchanged. The correct table can be found in McClellan and Parks, 1972, and is easily confirmed numerically.)

- F. A. Grünbaum (1982). "The eigenvectors of the discrete Fourier transform". *J. Math. Anal. Appl.* **88** (2): 355–363. doi:10.1016/0022-247X(82)90199-8.

- Natig M. Atakishiyev and Kurt Bernardo Wolf (1997). "Fractional Fourier-Kravchuk transform". *J. Opt. Soc. Am. A* **14** (7): 1467–1477. Bibcode:1997JOSAA..14.1467A. doi:10.1364/JOSAA.14.001467.

- C. Candan, M. A. Kutay and H. M.Ozaktas (2000). "The discrete fractional Fourier transform". *IEEE Trans. on Signal Processing* **48** (5): 1329–1337. Bibcode:2000ITSP...48.1329C. doi:10.1109/78.839980.

- Magdy Tawfik Hanna, Nabila Philip Attalla Seif, and Waleed Abd El Maguid Ahmed (2004). "Hermite-Gaussian-like eigenvectors of the discrete Fourier transform matrix based on the singular-value decomposition of its orthogonal projection matrices". *IEEE Trans. Circ. Syst. I* **51** (11): 2245–2254. doi:10.1109/TCSI.2004.836850.

- Shamgar Gurevich and Ronny Hadani (2009). "On the diagonalization of the discrete Fourier transform". *Applied and Computational Harmonic Analysis* **27** (1): 87–99. arXiv:0808.3281. doi:10.1016/j.acha.2008.11.003. preprint at.

- Shamgar Gurevich, Ronny Hadani, and Nir Sochen (2008). "The finite harmonic oscillator and its applications to sequences, communication and radar". *IEEE Transactions on Information Theory* **54** (9): 4239–4253. arXiv:0808.1495. doi:10.1109/TIT.2008.926440. preprint at.

- Juan G. Vargas-Rubio and Balu Santhanam (2005). "On the multiangle centered discrete fractional Fourier transform". *IEEE Sig. Proc. Lett.* **12** (4): 273–276. Bibcode:2005ISPL...12..273V. doi:10.1109/LSP.2005.843762.

- J. Cooley, P. Lewis, and P. Welch (1969). "The finite Fourier transform". *IEEE Trans. Audio Electroacoustics* **17** (2): 77–85. doi:10.1109/TAU.1969.1162036.

- F.N. Kong (2008). "Analytic Expressions of Two Discrete Hermite-Gaussian Signals". *IEEE Trans. Circuits and Systems –II: Express Briefs.* **55** (1): 56–60. doi:10.1109/TCSII.2007.909865.

6.13   External links

- Discrete Fourier Transform

- Interactive explanation of the DFT

- Matlab tutorial on the Discrete Fourier Transformation

- Interactive flash tutorial on the DFT

- Mathematics of the Discrete Fourier Transform by Julius O. Smith III

- Fast implementation of the DFT - coded in C and under General Public License (GPL)

- The DFT "à Pied": Mastering The Fourier Transform in One Day

- Explained: The Discrete Fourier Transform

- wavetable Cooker GPL application with graphical interface written in C, and implementing DFT IDFT to generate a wavetable set

- Online DFT coefficients calculator with graphical display – Gives online calculation of DFT coefficients with graphical display

- Discrete Fourier Transform JavaScript Program – A live, online, program for computing the discrete Fourier Transform coefficients of a real, uniformly-spaced, data sequence.

Chapter 7

List of Fourier-related transforms

This is a list of linear transformations of functions related to Fourier analysis. Such transformations map a function to a set of coefficients of basis functions, where the basis functions are sinusoidal and are therefore strongly localized in the frequency spectrum. (These transforms are generally designed to be invertible.) In the case of the Fourier transform, each basis function corresponds to a single frequency component.

7.1 Continuous transforms

Applied to functions of continuous arguments, Fourier-related transforms include:

- Two-sided Laplace transform

- Mellin transform, another closely related integral transform

- Laplace transform

- Fourier transform, with special cases:

  - Fourier series
    - When the input function/waveform is periodic, the Fourier transform output is a Dirac comb function, modulated by a discrete sequence of finite-valued coefficients that are complex-valued in general. These are called **Fourier series coefficients**. The term **Fourier series** actually refers to the inverse Fourier transform, which is a sum of sinusoids at discrete frequencies, weighted by the Fourier series coefficients.
    - When the non-zero portion of the input function has finite duration, the Fourier transform is continuous and finite-valued. But a discrete subset of its values is sufficient to reconstruct/represent the portion that was analyzed. The same discrete set is obtained by treating the duration of the segment as one period of a periodic function and computing the Fourier series coefficients.
  - Sine and cosine transforms: When the input function has odd or even symmetry around the origin, the Fourier transform reduces to a sine or cosine transform.

- Hartley transform

- Short-time Fourier transform (or short-term Fourier transform) (STFT)

  - Rectangular mask short-time Fourier transform

- Chirplet transform

- Fractional Fourier transform (FRFT)

- Hankel transform: related to the Fourier Transform of radial functions.

7.2   Discrete transforms

For usage on computers, number theory and algebra, discrete arguments (e.g. functions of a series of discrete samples) are often more appropriate, and are handled by the transforms (analogous to the continuous cases above):

- Discrete-time Fourier transform (DTFT): Equivalent to the Fourier transform of a "continuous" function that is constructed from the discrete input function by using the sample values to modulate a Dirac comb. When the sample values are derived by sampling a function on the real line, $f(x)$, the DTFT is equivalent to a periodic summation of the Fourier transform of f. The DTFT output is always periodic (cyclic). An alternative viewpoint is that the DTFT is a transform to a frequency domain that is bounded (or *finite*), the length of one cycle.

  - discrete Fourier transform (DFT):

    - When the input sequence is periodic, the DTFT output is also a Dirac comb function, modulated by the coefficients of a Fourier series[1] which can be computed as a DFT of one cycle of the input sequence. The number of discrete values in one cycle of the DFT is the same as in one cycle of the input sequence.

    - When the non-zero portion of the input sequence has finite duration, the DTFT is continuous and finite-valued. But a discrete subset of its values is sufficient to reconstruct/represent the portion that was analyzed. The same discrete set is obtained by treating the duration of the segment as one cycle of a periodic function and computing the DFT.

  - Discrete sine and cosine transforms: When the input sequence has odd or even symmetry around the origin, the DTFT reduces to a discrete sine transform (DST) or discrete cosine transform (DCT).

    - Regressive discrete Fourier series, in which the period is determined by the data rather than fixed in advance.

  - Discrete chebyshev transforms (on the 'roots' grid and the 'extrema' grid of the chebyshev polynomials of the first kind). This transform is of much importance in the field of spectral methods for solving differential equations because it can be used to swiftly and efficient go from grid point values to chebyshev series coefficients.

- Generalized DFT (GDFT), a generalization of the DFT and constant modulus transforms where phase functions might be of linear with integer and real valued slopes, or even non-linear phase bringing flexibilities for optimal designs of various metrics, e.g. auto- and cross-correlations.

- Z-transform, a generalization of the DTFT.

- Modified discrete cosine transform (MDCT)

- Discrete Hartley transform (DHT)

- Also the discretized STFT (see above).

- Hadamard transform (Walsh function).

The use of all of these transforms is greatly facilitated by the existence of efficient algorithms based on a fast Fourier transform (FFT). The Nyquist–Shannon sampling theorem is critical for understanding the output of such discrete transforms.

7.3   Notes

[1] The Fourier series represents $\sum_{n=-\infty}^{\infty} f(nT) \cdot \delta(t-nT)$, where T is the interval between samples.

7.4   See also

- Integral transform

- Wavelet transform

- Fourier transform spectroscopy

- Harmonic analysis

- List of transforms

- List of operators

- Bispectrum

7.5   References

- A. D. Polyanin and A. V. Manzhirov, *Handbook of Integral Equations*, CRC Press, Boca Raton, 1998. ISBN 0-8493-2876-4

- Tables of Integral Transforms at EqWorld: The World of Mathematical Equations.

- A.N. Akansu and H. Agirman-Tosun, "*Generalized Discrete Fourier Transform With Nonlinear Phase*", IEEE *Transactions on Signal Processing*, vol. 58, no. 9, pp. 4547-4556, Sept. 2010.

Chapter 8

Fast Fourier transform

"FFT" redirects here. For other uses, see FFT (disambiguation).

A **fast Fourier transform** (**FFT**) algorithm computes the discrete Fourier transform (DFT) of a sequence, or its inverse.

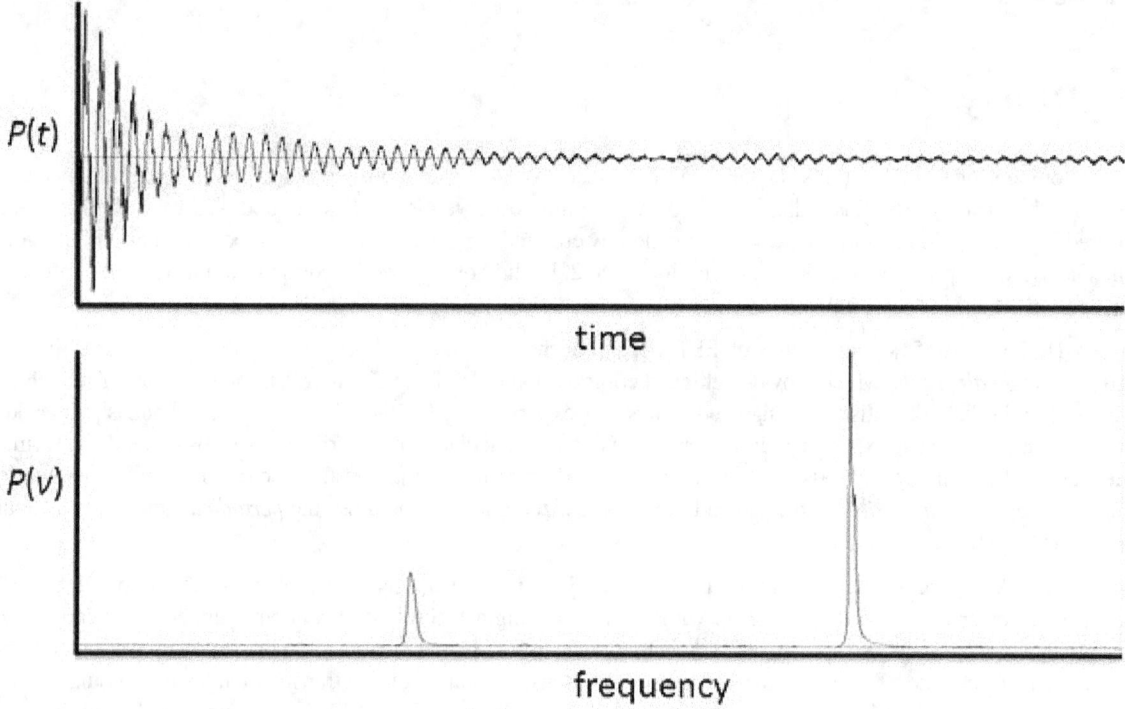

Time and frequency domain for the same signal

Fourier analysis converts a signal from its original domain (often time or space) to a representation in the frequency domain and vice versa. A FFT rapidly computes such transformations by factorizing the DFT matrix into a product of sparse (mostly zero) factors.[1] As a result, it manages to reduce the complexity of computing the DFT from $O(n^2)$, which arises if one simply applies the definition of DFT, to $O(n \log n)$, where n is the data size.

Fast Fourier transforms are widely used for many applications in engineering, science, and mathematics. The basic ideas were popularized in 1965, but some algorithms had been derived as early as 1805.[2] In 1994 Gilbert Strang described the FFT as "the most important numerical algorithm of our lifetime"[3] and it was included in Top 10 Algorithms of 20th Century by the IEEE journal Computing in Science & Engineering.[4]

8.1 Overview

There are many different FFT algorithms involving a wide range of mathematics, from simple complex-number arithmetic to group theory and number theory; this article gives an overview of the available techniques and some of their general properties, while the specific algorithms are described in subsidiary articles linked below.

The DFT is obtained by decomposing a sequence of values into components of different frequencies. This operation is useful in many fields (see discrete Fourier transform for properties and applications of the transform) but computing it directly from the definition is often too slow to be practical. An FFT is a way to compute the same result more quickly: computing the DFT of N points in the naive way, using the definition, takes $O(N^2)$ arithmetical operations, while an FFT can compute the same DFT in only $O(N \log N)$ operations. The difference in speed can be enormous, especially for long data sets where N may be in the thousands or millions. In practice, the computation time can be reduced by several orders of magnitude in such cases, and the improvement is roughly proportional to $N / \log(N)$. This huge improvement made the calculation of the DFT practical; FFTs are of great importance to a wide variety of applications, from digital signal processing and solving partial differential equations to algorithms for quick multiplication of large integers.

The best-known FFT algorithms depend upon the factorization of N, but there are FFTs with $O(N \log N)$ complexity for all N, even for prime N. Many FFT algorithms only depend on the fact that $e^{-\frac{2\pi i}{N}}$ is an N-th primitive root of unity, and thus can be applied to analogous transforms over any finite field, such as number-theoretic transforms. Since the inverse DFT is the same as the DFT, but with the opposite sign in the exponent and a $1/N$ factor, any FFT algorithm can easily be adapted for it.

8.2 History

The development of fast algorithms for DFT can be traced to Gauss's unpublished work in 1805 when he needed it to interpolate the orbit of asteroids Pallas and Juno from sample observations.[5] His method was very similar to the one published in 1965 by Cooley and Tukey, who are generally credited for the invention of the modern generic FFT algorithm. While Gauss's work predated even Fourier's results in 1822, he did not analyze the computation time and eventually used other methods to achieve his goal.

Between 1805 and 1965, some versions of FFT were published by other authors. Yates in 1932 published his version called *interaction algorithm*, which provided efficient computation of Hadamard and Walsh transforms.[6] Yates' algorithm is still used in the field of statistical design and analysis of experiments. In 1942, Danielson and Lanczos published their version to compute DFT for x-ray crystallography, a field where calculation of Fourier transforms presented a formidable bottleneck.[7] While many methods in the past had focused on reducing the constant factor for $O(n^2)$ computation by taking advantage of *symmetries*, Danielson and Lanczos realized that one could use the *periodicity* and apply a "doubling trick" to get $O(n \log n)$ runtime.[8]

Cooley and Tukey published a more general version of FFT in 1965 that is applicable when N is composite and not necessarily a power of 2.[9] Tukey came up with the idea during a meeting of President Kennedy's Science Advisory Committee where a discussion topic involved detecting nuclear tests by the Soviet Union by setting up sensors to surround the country from outside. To analyze the output of these sensors, a fast Fourier transform algorithm would be needed. Tukey's idea was taken by Richard Garwin and given to Cooley (both worked at IBM's Watson labs) for implementation while hiding the original purpose from him for security reasons. The pair published the paper in a relatively short six months.[10] As Tukey didn't work at IBM, the patentability of the idea was doubted and the algorithm went into the public domain, which, through the computing revolution of the next decade, made FFT one of the indispensable algorithms in digital signal processing.

8.3 Definition and speed

An FFT computes the DFT and produces exactly the same result as evaluating the DFT definition directly; the most important difference is that an FFT is much faster. (In the presence of round-off error, many FFT algorithms are also much more accurate than evaluating the DFT definition directly, as discussed below.)

Let $x_0, ..., x_{N-1}$ be complex numbers. The DFT is defined by the formula

$$X_k = \sum_{n=0}^{N-1} x_n e^{-i2\pi k \frac{n}{N}} \qquad k = 0, \ldots, N-1.$$

Evaluating this definition directly requires $O(N^2)$ operations: there are N outputs X_k, and each output requires a sum of N terms. An FFT is any method to compute the same results in $O(N \log N)$ operations. More precisely, all known FFT algorithms require $\Theta(N \log N)$ operations (technically, O only denotes an upper bound), although there is no known proof that a lower complexity score is impossible.(Johnson and Frigo, 2007)

To illustrate the savings of an FFT, consider the count of complex multiplications and additions. Evaluating the DFT's sums directly involves N^2 complex multiplications and $N(N-1)$ complex additions [of which $O(N)$ operations can be saved by eliminating trivial operations such as multiplications by 1]. The well-known radix-2 Cooley–Tukey algorithm, for N a power of 2, can compute the same result with only $(N/2)\log_2(N)$ complex multiplications (again, ignoring simplifications of multiplications by 1 and similar) and $N\log_2(N)$ complex additions. In practice, actual performance on modern computers is usually dominated by factors other than the speed of arithmetic operations and the analysis is a complicated subject (see, e.g., Frigo & Johnson, 2005), but the overall improvement from $O(N^2)$ to $O(N \log N)$ remains.

8.4   Algorithms

8.4.1   Cooley–Tukey algorithm

Main article: Cooley–Tukey FFT algorithm

By far the most commonly used FFT is the Cooley–Tukey algorithm. This is a divide and conquer algorithm that recursively breaks down a DFT of any composite size $N = N_1N_2$ into many smaller DFTs of sizes N_1 and N_2, along with $O(N)$ multiplications by complex roots of unity traditionally called twiddle factors (after Gentleman and Sande, 1966[11]).

This method (and the general idea of an FFT) was popularized by a publication of J. W. Cooley and J. W. Tukey in 1965,[9] but it was later discovered[2] that those two authors had independently re-invented an algorithm known to Carl Friedrich Gauss around 1805[12] (and subsequently rediscovered several times in limited forms).

The best known use of the Cooley–Tukey algorithm is to divide the transform into two pieces of size $N/2$ at each step, and is therefore limited to power-of-two sizes, but any factorization can be used in general (as was known to both Gauss and Cooley/Tukey[2]). These are called the **radix-2** and **mixed-radix** cases, respectively (and other variants such as the split-radix FFT have their own names as well). Although the basic idea is recursive, most traditional implementations rearrange the algorithm to avoid explicit recursion. Also, because the Cooley–Tukey algorithm breaks the DFT into smaller DFTs, it can be combined arbitrarily with any other algorithm for the DFT, such as those described below.

8.4.2   Other FFT algorithms

Main articles: Prime-factor FFT algorithm, Bruun's FFT algorithm, Rader's FFT algorithm and Bluestein's FFT algorithm

There are other FFT algorithms distinct from Cooley–Tukey.

Cornelius Lanczos did pioneering work on the FFS and FFT with G.C. Danielson (1940).

For $N = N_1N_2$ with coprime N_1 and N_2, one can use the Prime-Factor (Good-Thomas) algorithm (PFA), based on the Chinese Remainder Theorem, to factorize the DFT similarly to Cooley–Tukey but without the twiddle factors. The Rader-Brenner algorithm (1976) is a Cooley–Tukey-like factorization but with purely imaginary twiddle factors, reducing multiplications at the cost of increased additions and reduced numerical stability; it was later superseded by the split-radix variant of Cooley–Tukey (which achieves the same multiplication count but with fewer additions and without sacrificing accuracy). Algorithms that recursively factorize the DFT into smaller operations other than DFTs include the Bruun and QFT algorithms. (The Rader-Brenner and QFT algorithms were proposed for power-of-two sizes, but it is possible that they could be adapted to general composite n. Bruun's algorithm applies to arbitrary even composite sizes.) Bruun's algorithm, in particular, is based on interpreting the FFT as a recursive factorization of the polynomial z^N-1, here into real-coefficient polynomials of the form z^M-1 and $z^{2M} + az^M + 1$.

Another polynomial viewpoint is exploited by the Winograd algorithm, which factorizes z^N-1 into cyclotomic polynomials—these often have coefficients of 1, 0, or −1, and therefore require few (if any) multiplications, so Winograd can be used to obtain minimal-multiplication FFTs and is often used to find efficient algorithms for small factors. Indeed, Winograd showed that the DFT can be computed with only $O(N)$ irrational multiplications, leading to a proven achievable lower bound on the number of multiplications for power-of-two sizes; unfortunately, this comes at the cost of many more additions, a tradeoff no longer favorable on modern processors with hardware multipliers. In particular, Winograd also makes use of the PFA as well as an algorithm by Rader for FFTs of *prime* sizes.

Rader's algorithm, exploiting the existence of a generator for the multiplicative group modulo prime N, expresses a DFT of prime size n as a cyclic convolution of (composite) size $N-1$, which can then be computed by a pair of ordinary FFTs via the convolution theorem (although Winograd uses other convolution methods). Another prime-size FFT is due to L. I. Bluestein, and is sometimes called the chirp-z algorithm; it also re-expresses a DFT as a convolution, but this time of the *same* size (which can be zero-padded to a power of two and evaluated by radix-2 Cooley–Tukey FFTs, for example), via the identity $nk = -(k-n)^2/2 + n^2/2 + k^2/2$.

8.5 FFT algorithms specialized for real and/or symmetric data

In many applications, the input data for the DFT are purely real, in which case the outputs satisfy the symmetry

$$X_{N-k} = X_k^*$$

and efficient FFT algorithms have been designed for this situation (see e.g. Sorensen, 1987). One approach consists of taking an ordinary algorithm (e.g. Cooley–Tukey) and removing the redundant parts of the computation, saving roughly a factor of two in time and memory. Alternatively, it is possible to express an *even*-length real-input DFT as a complex DFT of half the length (whose real and imaginary parts are the even/odd elements of the original real data), followed by $O(N)$ post-processing operations.

It was once believed that real-input DFTs could be more efficiently computed by means of the discrete Hartley transform (DHT), but it was subsequently argued that a specialized real-input DFT algorithm (FFT) can typically be found that requires fewer operations than the corresponding DHT algorithm (FHT) for the same number of inputs. Bruun's algorithm (above) is another method that was initially proposed to take advantage of real inputs, but it has not proved popular.

There are further FFT specializations for the cases of real data that have even/odd symmetry, in which case one can gain another factor of (roughly) two in time and memory and the DFT becomes the discrete cosine/sine transform(s) (DCT/DST). Instead of directly modifying an FFT algorithm for these cases, DCTs/DSTs can also be computed via FFTs of real data combined with $O(N)$ pre/post processing.

8.6 Computational issues

8.6.1 Bounds on complexity and operation counts

A fundamental question of longstanding theoretical interest is to prove lower bounds on the complexity and exact operation counts of fast Fourier transforms, and many open problems remain. It is not even rigorously proved whether DFTs truly require $\Omega(N \log(N))$ (i.e., order $N \log(N)$ or greater) operations, even for the simple case of power of two sizes, although no algorithms with lower complexity are known. In particular, the count of arithmetic operations is usually the focus of such questions, although actual performance on modern-day computers is determined by many other factors such as cache or CPU pipeline optimization.

Following pioneering work by Winograd (1978), a tight $\Theta(N)$ lower bound *is* known for the number of real multiplications required by an FFT. It can be shown that only $4N - 2\log_2^2 N - 2\log_2 N - 4$ irrational real multiplications are required to compute a DFT of power-of-two length $N = 2^m$. Moreover, explicit algorithms that achieve this count are known (Heideman & Burrus, 1986; Duhamel, 1990). Unfortunately, these algorithms require too many additions to be practical, at least on modern computers with hardware multipliers (Duhamel, 1990; Frigo & Johnson, 2005).

A tight lower bound is *not* known on the number of required additions, although lower bounds have been proved under some restrictive assumptions on the algorithms. In 1973, Morgenstern proved an $\Omega(N \log(N))$ lower bound on the addition count for algorithms where the multiplicative constants have bounded magnitudes (which is true for most but not all FFT algorithms). This result, however, applies only to the unnormalized Fourier transform (which is a scaling of a unitary matrix by a factor of $\sqrt{N}$), and does not explain why the Fourier matrix is harder to compute than any other unitary matrix (including the identity matrix) under the same scaling. Pan (1986) proved an $\Omega(N \log(N))$ lower bound assuming a bound on a measure of the FFT algorithm's "asynchronicity", but the generality of this assumption is unclear. For the case of power-of-two N, Papadimitriou (1979) argued that the number $N \log_2 N$ of complex-number additions achieved by Cooley–Tukey algorithms is *optimal* under certain assumptions on the graph of the algorithm (his assumptions imply, among other things, that no additive identities in the roots of unity are exploited). (This argument would imply that at least $2N \log_2 N$ real additions are required, although this is not a tight bound because extra additions are required as part of complex-number multiplications.) Thus far, no published FFT algorithm has achieved fewer than $N \log_2 N$ complex-number additions (or their equivalent) for power-of-two N.

A third problem is to minimize the *total* number of real multiplications and additions, sometimes called the "arithmetic complexity" (although in this context it is the exact count and not the asymptotic complexity that is being considered). Again, no tight lower bound has been proven. Since 1968, however, the lowest published count for power-of-two N was long achieved by the split-radix FFT algorithm, which requires $4N \log_2 N - 6N + 8$ real multiplications and additions for $N > 1$. This was recently reduced to $\sim \frac{34}{9} N \log_2 N$ (Johnson and Frigo, 2007; Lundy and Van Buskirk, 2007). A slightly larger count (but still better than split radix for $N \geq 256$) was shown to be provably optimal for $N \leq 512$ under additional restrictions on the possible algorithms (split-radix-like flowgraphs with unit-modulus multiplicative factors), by reduction to a Satisfiability Modulo Theories problem solvable by brute force (Haynal & Haynal, 2011).

Most of the attempts to lower or prove the complexity of FFT algorithms have focused on the ordinary complex-data case, because it is the simplest. However, complex-data FFTs are so closely related to algorithms for related problems such as real-data FFTs, discrete cosine transforms, discrete Hartley transforms, and so on, that any improvement in one of these would immediately lead to improvements in the others (Duhamel & Vetterli, 1990).

8.6.2   Approximations

All of the FFT algorithms discussed above compute the DFT exactly (in exact arithmetic, i.e. neglecting floating-point errors). A few "FFT" algorithms have been proposed, however, that compute the DFT *approximately*, with an error that can be made arbitrarily small at the expense of increased computations. Such algorithms trade the approximation error for increased speed or other properties. For example, an approximate FFT algorithm by Edelman et al. (1999) achieves lower communication requirements for parallel computing with the help of a fast multipole method. A wavelet-based approximate FFT by Guo and Burrus (1996) takes sparse inputs/outputs (time/frequency localization) into account more efficiently than is possible with an exact FFT. Another algorithm for approximate computation of a subset of the DFT outputs is due to Shentov et al. (1995). The Edelman algorithm works equally well for sparse and non-sparse data, since it is based on the compressibility (rank deficiency) of the Fourier matrix itself rather than the compressibility (sparsity) of the data. Conversely, if the data are sparse—that is, if only K out of N Fourier coefficients are nonzero—then the complexity can be reduced to $O(K\log(N)\log(N/K))$, and this has been demonstrated to lead to practical speedups compared to an ordinary FFT for $N/K > 32$ in a large-N example ($N=2^{22}$) using a probabilistic approximate algorithm (which estimates the largest K coefficients to several decimal places).[13]

8.6.3 Accuracy

Even the "exact" FFT algorithms have errors when finite-precision floating-point arithmetic is used, but these errors are typically quite small; most FFT algorithms, e.g. Cooley–Tukey, have excellent numerical properties as a consequence of the pairwise summation structure of the algorithms. The upper bound on the relative error for the Cooley–Tukey algorithm is $O(\varepsilon \log N)$, compared to $O(\varepsilon N^{3/2})$ for the naïve DFT formula,[11] where ε is the machine floating-point relative precision. In fact, the root mean square (rms) errors are much better than these upper bounds, being only $O(\varepsilon \sqrt{\log N})$ for Cooley–Tukey and $O(\varepsilon \sqrt{N})$ for the naïve DFT (Schatzman, 1996). These results, however, are very sensitive to the accuracy of the twiddle factors used in the FFT (i.e. the trigonometric function values), and it is not unusual for incautious FFT implementations to have much worse accuracy, e.g. if they use inaccurate trigonometric recurrence formulas. Some FFTs other than Cooley–Tukey, such as the Rader-Brenner algorithm, are intrinsically less stable.

In fixed-point arithmetic, the finite-precision errors accumulated by FFT algorithms are worse, with rms errors growing as $O(\sqrt{N})$ for the Cooley–Tukey algorithm (Welch, 1969). Moreover, even achieving this accuracy requires careful attention to scaling to minimize loss of precision, and fixed-point FFT algorithms involve rescaling at each intermediate stage of decompositions like Cooley–Tukey.

To verify the correctness of an FFT implementation, rigorous guarantees can be obtained in $O(N\log(N))$ time by a simple procedure checking the linearity, impulse-response, and time-shift properties of the transform on random inputs (Ergün, 1995).

8.7   Multidimensional FFTs

As defined in the multidimensional DFT article, the multidimensional DFT

$$X_{\mathbf{k}} = \sum_{\mathbf{n}=0}^{\mathbf{N}-1} e^{-2\pi i \mathbf{k} \cdot (\mathbf{n}/\mathbf{N})} x_{\mathbf{n}}$$

transforms an array $x_{\mathbf{n}}$ with a d-dimensional vector of indices $\mathbf{n} = (n_1, \ldots, n_d)$ by a set of d nested summations (over $n_j = 0 \ldots N_j - 1$ for each j), where the division $\mathbf{n}/\mathbf{N}$, defined as $\mathbf{n}/\mathbf{N} = (n_1/N_1, \ldots, n_d/N_d)$, is performed element-wise. Equivalently, it is the composition of a sequence of d sets of one-dimensional DFTs, performed along one dimension at a time (in any order).

This compositional viewpoint immediately provides the simplest and most common multidimensional DFT algorithm, known as the **row-column** algorithm (after the two-dimensional case, below). That is, one simply performs a sequence of d one-dimensional FFTs (by any of the above algorithms): first you transform along the n_1 dimension, then along the n_2 dimension, and so on (or actually, any ordering works). This method is easily shown to have the usual $O(N\log(N))$ complexity, where $N = N_1 \cdot N_2 \cdot \ldots \cdot N_d$ is the total number of data points transformed. In particular, there are N/N_1 transforms of size N_1, etcetera, so the complexity of the sequence of FFTs is:

$$\frac{N}{N_1} O(N_1 \log N_1) + \cdots + \frac{N}{N_d} O(N_d \log N_d)$$

$$= O\left(N\left[\log N_1 + \cdots + \log N_d\right]\right) = O(N \log N).$$

In two dimensions, the $x_{\mathbf{k}}$ can be viewed as an $n_1 \times n_2$ matrix, and this algorithm corresponds to first performing the FFT of all the rows (resp. columns), grouping the resulting transformed rows (resp. columns) together as another $n_1 \times n_2$ matrix, and then performing the FFT on each of the columns (resp. rows) of this second matrix, and similarly grouping the results into the final result matrix.

In more than two dimensions, it is often advantageous for cache locality to group the dimensions recursively. For example, a three-dimensional FFT might first perform two-dimensional FFTs of each planar "slice" for each fixed n_1, and then perform the one-dimensional FFTs along the n_1 direction. More generally, an asymptotically optimal cache-oblivious algorithm consists of recursively dividing the dimensions into two groups $(n_1, \ldots, n_{d/2})$ and $(n_{d/2+1}, \ldots, n_d)$ that are transformed recursively (rounding if d is not even) (see Frigo and Johnson, 2005). Still, this remains a straightforward variation of the row-column algorithm that ultimately requires only a one-dimensional FFT algorithm as the base case, and still has O(Nlog(N)) complexity. Yet another variation is to perform matrix transpositions in between transforming subsequent dimensions, so that the transforms operate on contiguous data; this is especially important for out-of-core and distributed memory situations where accessing non-contiguous data is extremely time-consuming.

There are other multidimensional FFT algorithms that are distinct from the row-column algorithm, although all of them have O(Nlog(N)) complexity. Perhaps the simplest non-row-column FFT is the vector-radix FFT algorithm, which is a generalization of the ordinary Cooley–Tukey algorithm where one divides the transform dimensions by a vector $\mathbf{r} = (r_1, r_2, \ldots, r_d)$ of radices at each step. (This may also have cache benefits.) The simplest case of vector-radix is where all of the radices are equal (e.g. vector-radix-2 divides *all* of the dimensions by two), but this is not necessary. Vector radix with only a single non-unit radix at a time, i.e. $\mathbf{r} = (1, \ldots, 1, r, 1, \ldots, 1)$, is essentially a row-column algorithm. Other, more complicated, methods include polynomial transform algorithms due to Nussbaumer (1977), which view the transform in terms of convolutions and polynomial products. See Duhamel and Vetterli (1990) for more information and references.

8.8   Other generalizations

An O($N^{5/2}$log(N)) generalization to spherical harmonics on the sphere S^2 with N^2 nodes was described by Mohlenkamp,[14] along with an algorithm conjectured (but not proven) to have O($N^2 \log^2(N)$) complexity; Mohlenkamp also provides an implementation in the libftsh library. A spherical-harmonic algorithm with O(N^2log(N)) complexity is described by Rokhlin and Tygert.[15]

The Fast Folding Algorithm is analogous to the FFT, except that it operates on a series of binned waveforms rather than a series of real or complex scalar values. Rotation (which in the FFT is multiplication by a complex phasor) is a circular shift of the component waveform.

Various groups have also published "FFT" algorithms for non-equispaced data, as reviewed in Potts *et al.* (2001). Such algorithms do not strictly compute the DFT (which is only defined for equispaced data), but rather some approximation thereof (a non-uniform discrete Fourier transform, or NDFT, which itself is often computed only approximately). More generally there are various other methods of spectral estimation.

8.9   Applications

FFT's importance derives from the fact that in signal processing and image processing it has made working in frequency domain equally computationally feasible as working in temporal or spatial domain. Some of the important applications of FFT includes,[10][16]

- Fast large integer and polynomial multiplication

- Efficient matrix-vector multiplication for Toeplitz, circulant and other structured matrices

- Filtering algorithms

- Fast algorithms for discrete cosine or sine transforms (example, Fast DCT used for JPEG, MP3/MPEG encoding)

- Fast Chebyshev approximation

- Fast Discrete Hartley Transform

- Solving Difference Equations

8.10 Time-Frequency Relationships

[Note: the terms "complex data" (= real and imaginary numbers) and "in-phase/quadrature data" (I/Q numbers) are used interchangeably].

8.10.1 Time Domain

The waveform is sampled at **N** equi-spaced points in time $n = 0, 1, ..., N - 1$
(It is preferable for N to be a power of two).
When the sampled waveform is real, the data consists of N real numbers—but if the sampled waveform is complex, the data consists of N sample-pairs of I/Q data.
If the time interval between samples (the "sample interval") is ▯t secs, then
the total length of the sample record is **T** , where

T = N.▯t secs

8.10.2 Frequency Domain

For real waveform data, the number of frequency points, ignoring the zero frequency bin, is N/2 +1. [i.e.there are N/2 +1 pairs of (I/Q) data].
For complex (I/Q) waveform data, the number of frequency points = N, with N/2 pairs of complex data at positive frequencies and N/2 pairs of complex data at negative frequencies.
If the frequency separation of the data points, (the "frequency resolution") is ▯f Hz, then
the maximum frequency of the display (the "bandwidth") is $+F_{max}$, for real numbers, and $\pm F_{max}$ for complex data, where

$\mathbf{F}_{max} = \frac{N}{2}.\mathbf{▯f}$ Hz

8.10.3 Interrelationships between time and frequency domains

▯t $= \frac{1}{2.F_{max}}$ secs
▯f $= \frac{1}{T}$ Hz

8.11 Research Areas

- **Big FFTs**: With explosion of big data in fields such as astronomy, the need for 512k FFTs has arisen for certain interferometry calculations. The data collected by projects such as MAP and LIGO require FFTs of tens of billions of points. As this size does not fit in to main memory, so called out-of-core FFTs are an active area of research.[17]

- **Approximate FFTs**: For applications such as MRI, it is necessary to compute DFTs for nonuniformly spaced grid points and/or frequencies. Multipole based approaches can compute approximate quantities with factor of runtime increase.[18]

- **Group FFTs**: The FFT may also be explained and interpreted using group representation theory that allows for further generalization. A function on any compact group, including non cyclic, has an expansion in terms of a basis of irreducible matrix elements. It remains active area of research to find efficient algorithm for performing this change of basis. Applications including efficient spherical harmonic expansion, analyzing certain markov processes, robotics etc.[19]

- **Quantum FFTs**: Shor's fast algorithm for integer factorization on a quantum computer has a subroutine to compute DFT of a binary vector. This is implemented as sequence of 1- or 2-bit quantum gates now known as quantum FFT, which is effectively the Cooley-Tukey FFT realized as a particular factorization of the Fourier matrix. Extension to these ideas is currently being explored.

8.12   Implementation with C++

Here is a C++ source code for a simple FFT implementation:[20]

```
const double TwoPi = 6.283185307179586; void FFTAnalysis(double *AVal, double *FTvl, int Nvl, int Nft) { int i, j, n,
m, Mmax, Istp; double Tmpr, Tmpi, Wtmp, Theta; double Wpr, Wpi, Wr, Wi; double *Tmvl; n = Nvl * 2; Tmvl = new
double[n+1]; for (i = 0; i <Nvl; i++) { j = i * 2; Tmvl[j] = 0; Tmvl[j+1] = AVal[i]; } i = 1; j = 1; while (i <n) { if (j> i) {
Tmpr = Tmvl[i]; Tmvl[i] = Tmvl[j]; Tmvl[j] = Tmpr; Tmpr = Tmvl[i+1]; Tmvl[i+1] = Tmvl[j+1]; Tmvl[j+1] = Tmpr;
} i = i + 2; m = Nvl; while ((m>= 2) && (j> m)) { j = j - m; m = m>> 2; } j = j + m; } Mmax = 2; while (n> Mmax)
{ Theta = -TwoPi / Mmax; Wpi = Sin(Theta); Wtmp = Sin(Theta / 2); Wpr = Wtmp * Wtmp * 2; Istp = Mmax * 2; Wr
= 1; Wi = 0; m = 1; while (m <Mmax) { i = m; m = m + 2; Tmpr = Wr; Tmpi = Wi; Wr = Wr - Tmpr * Wpr - Tmpi
* Wpi; Wi = Wi + Tmpr * Wpi - Tmpi * Wpr; while (i <n) { j = i + Mmax; Tmpr = Wr * Tmvl[j] - Wi * Tmvl[j+1];
Tmpi = Wi * Tmvl[j] + Wr * Tmvl[j+1]; Tmvl[j] = Tmvl[i] - Tmpr; Tmvl[j+1] = Tmvl[i+1] - Tmpi; Tmvl[i] = Tmvl[i]
+ Tmpr; Tmvl[i+1] = Tmvl[i+1] + Tmpi; i = i + Istp; } } Mmax = Istp; } for (i = 1; i <Nft; i++) { j = i * 2; FTvl[i] =
Sqrt(Sqr(Tmvl[j]) + Sqr(Tmvl[j+1])); } delete []Tmvl; }
```

8.13   Meaning of 'negative frequencies' from fast Fourier transforming real valued data

The majority of FFT algorithms result in 'negative frequencies' from Fourier transforming real numbers. The negative frequencies encompass half of the resulting numbers in the transform and provide no new information, since they are conjugate pairs of the positive frequencies. In contrast, Fourier transforming numbers with real and imaginary components does not result in a transformed set of numbers with such a redundancy.

The meaning of these negative frequencies resulting from the Fourier transform of real valued data can be intuited through the DFT matrix. Briefly, the DFT matrix performs a Fourier transform on a set of numbers as a matrix multiplication. The matrix works based on the orthogonality of sinusoidal functions, the real-world analogue of which is a lock-in amplifier. When a sinusoidal function of frequency f1 is multiplied by another sinusoidal function of frequency f2 not equal to f1 and integrated, the result is zero. Instead, when f1 is equal to f2, the average value is equal to half of the product of the amplitudes. The DFT matrix is a matrix of sinusoidal frequencies with an amplitude of 1, with the first row being 0Hz, up to the final row being a sinusoid with N-1 periods (the fastest encodeable frequency with slightly more than one datapoint per period). The phase of f2 is encoded in the real and imaginary component of the transformed value.

That the DFT matrix transforms frequencies with less than 2 samples per period might be viewed as being in violation of the Nyquist–Shannon sampling theorem, which demands that at least two samples per period are necessary to resolve frequencies from a dataset. While this requirement is true for real-valued data which encodes amplitude at every point, complex valued data encodes both amplitude and phase at every point. This two-part complex value conveys twice as much information as a single real-valued datapoint, and as little as one sample per period is necessary to unambiguously resolve frequencies from a complex valued array of numbers.

Despite that real-valued data cannot resolve frequencies with less than 2 samples per period, common Fourier transform algorithms (and the DFT matrix) will still attempt to resolve these frequencies. The result of trying to recover frequencies with less than 2 real valued datapoints is that nonsense 'negative frequencies' are generated in the transform. For example, a sine wave sampled 8 times per period results in a dataset that is identical to a sine wave sampled 8/7 times per period. The Fourier transform will attempt to resolve both of these frequencies, and attempting to recover a waveform sampled 8/7 times per period results in a 'negative frequency' conjugate.

8.14 See also

- Cooley–Tukey FFT algorithm

- Prime-factor FFT algorithm

- Bruun's FFT algorithm

- Rader's FFT algorithm

- Bluestein's FFT algorithm

- Butterfly diagram – a diagram used to describe FFTs.

- Odlyzko–Schönhage algorithm applies the FFT to finite Dirichlet series.

- Overlap add/Overlap save – efficient convolution methods using FFT for long signals

- Spectral music (involves application of FFT analysis to musical composition)

- Spectrum analyzers – Devices that perform an FFT

- FFTW "Fastest Fourier Transform in the West" – C library for the discrete Fourier transform (DFT) in one or more dimensions.

- FFTS – The Fastest Fourier Transform in the South.

- FFTPACK – another Fortran FFT library (public domain)

- Goertzel algorithm – Computes individual terms of discrete Fourier transform

- Time series

- Math Kernel Library

- Fast Walsh–Hadamard transform

- Generalized distributive law

- Multidimensional transform

- Multidimensional discrete convolution

8.15 References

[1] Charles Van Loan, *Computational Frameworks for the Fast Fourier Transform* (SIAM, 1992).

[2] Heideman, M. T.; Johnson, D. H.; Burrus, C. S. (1984). "Gauss and the history of the fast Fourier transform". *IEEE ASSP Magazine* **1** (4): 14–21. doi:10.1109/MASSP.1984.1162257.

[3] Strang, Gilbert (May–June 1994). "Wavelets". *American Scientist* **82** (3): 253. JSTOR 29775194.

[4] Dongarra, J.; Sullivan, F. (January 2000). "Guest Editors Introduction to the top 10 algorithms". *Computing in Science Engineering* **2** (1): 22–23. doi:10.1109/MCISE.2000.814652. ISSN 1521-9615.

[5] Heideman, Michael T.; Johnson, Don H.; Burrus, C. Sidney (1985-09-01). "Gauss and the history of the fast Fourier transform". *Archive for History of Exact Sciences* **34** (3): 265–277. doi:10.1007/BF00348431. ISSN 0003-9519.

[6] Yates, Frank (1937). "The design and analysis of factorial experiments". *Technical Communication no. 35 of the Commonwealth Bureau of Soils.*

[7] Danielson and Lanczos (1942). "Some improvements in practical Fourier analysis and their application to x-ray scattering from liquids". *Journal of the Franklin Institute* **233** (4): 365–380. doi:10.1016/S0016-0032(42)90767-1.

[8] Cooley, James W.; Lewis, Peter A.W.; Welch, Peter D. (June 1967). "Historical notes on the fast Fourier transform". *IEEE Transactions on Audio and Electroacoustics* **15** (2): 76–79. doi:10.1109/TAU.1967.1161903. ISSN 0018-9278.

[9] Cooley, James W.; Tukey, John W. (1965). "An algorithm for the machine calculation of complex Fourier series". *Mathematics of Computation* **19** (90): 297–301. doi:10.1090/S0025-5718-1965-0178586-1. ISSN 0025-5718.

[10] Rockmore, D.N. (January 2000). "The FFT: an algorithm the whole family can use". *Computing in Science Engineering* **2** (1): 60–64. doi:10.1109/5992.814659. ISSN 1521-9615.

[11] Gentleman, W. M.; Sande, G. (1966). "Fast Fourier transforms—for fun and profit". *Proc. AFIPS* **29**: 563–578. doi:10.1145/14

[12] Carl Friedrich Gauss, 1866. "Theoria interpolationis methodo nova tractata," *Werke* band **3**, 265–327. Göttingen: Königliche Gesellschaft der Wissenschaften.

[13] Haitham Hassanieh, Piotr Indyk, Dina Katabi, and Eric Price, "Simple and Practical Algorithm for Sparse Fourier Transform" (PDF), ACM-SIAM Symposium On Discrete Algorithms (SODA), Kyoto, January 2012. See also the sFFT Web Page.

[14] Mohlenkamp, Martin J. (1999). "A F ast T ransform for Spherical Harmonics" (PDF). *The Journal of F ourier A nalysis and Applic ations* **5** (2/3): 159–184. Retrieved 18 September 2014.

[15] ROKHLIN, VLADIMIR; TYGERT, MARK (2006). "FAST ALGORITHMS FOR SPHERICAL HARMONIC EXPAN-SIONS" (PDF). *SIAM Journal on Scientific Computing* **27** (6): 903–1928. Retrieved 18 September 2014.

[16] Chu and George. "16". *Inside the FFT Black Box: Serial and Parallel Fast Fourier Transform Algorithms*. CRC Press. pp. 153–168. ISBN 9781420049961.

[17] Cormen and Nicol (1998). "Performing out-of-core FFTs on parallel disk systems". *Parallel Computing* **24**(1): 5–20. doi:10.101 8191(97)00114-2.

[18] Dutt, A.; Rokhlin, V. (November 1, 1993). "Fast Fourier Transforms for Nonequispaced Data". *SIAM Journal on Scientific Computing* **14** (6): 1368–1393. doi:10.1137/0914081. ISSN 1064-8275.

[19] Rockmore, Daniel N. (2004). Byrnes, Jim, ed. "Recent Progress and Applications in Group FFTs". *Computational Noncommutative Algebra and Applications*. NATO Science Series II: Mathematics, Physics and Chemistry (Springer Netherlands) **136**: 227–254. ISBN 978-1-4020-1982-1.

[20] fa:هیروف عیرس لیدبت

- Brenner, N.; Rader, C. (1976). "A New Principle for Fast Fourier Transformation". *IEEE Acoustics, Speech & Signal Processing* **24** (3): 264–266. doi:10.1109/TASSP.1976.1162805.

- Brigham, E. O. (2002). "The Fast Fourier Transform". New York: Prentice-Hall

- Thomas H. Cormen, Charles E. Leiserson, Ronald L. Rivest, and Clifford Stein, 2001. *Introduction to Algorithms*, 2nd. ed. MIT Press and McGraw-Hill. ISBN 0-262-03293-7. Especially chapter 30, "Polynomials and the FFT."

- Duhamel, Pierre (1990). "Algorithms meeting the lower bounds on the multiplicative complexity of length-2^n DFTs and their connection with practical algorithms". *IEEE Trans. Acoust. Speech. Sig. Proc.* **38** (9): 1504–151. doi:10.1109/29.60070.

- P. Duhamel and M. Vetterli, 1990, Fast Fourier transforms: a tutorial review and a state of the art, *Signal Processing* **19**: 259–299.

- A. Edelman, P. McCorquodale, and S. Toledo, 1999, The Future Fast Fourier Transform?, *SIAM J. Sci. Computing* **20**: 1094–1114.

- D. F. Elliott, & K. R. Rao, 1982, *Fast transforms: Algorithms, analyses, applications*. New York: Academic Press.

- Funda Ergün, 1995, Testing multivariate linear functions: Overcoming the generator bottleneck, *Proc. 27th ACM Symposium on the Theory of Computing*: 407–416.

- Frigo, M.; Johnson, S. G. (2005). "The Design and Implementation of FFTW3" (PDF). *Proceedings of the IEEE* **93**: 216–231. doi:10.1109/jproc.2004.840301.

- H. Guo and C. S. Burrus, 1996, Fast approximate Fourier transform via wavelets transform, *Proc. SPIE Intl. Soc. Opt. Eng.* **2825**: 250–259.

- H. Guo, G. A. Sitton, C. S. Burrus, 1994, The Quick Discrete Fourier Transform, *Proc. IEEE Conf. Acoust. Speech and Sig. Processing (ICASSP)* **3**: 445–448.

- Steve Haynal and Heidi Haynal, "Generating and Searching Families of FFT Algorithms", *Journal on Satisfiability, Boolean Modeling and Computation* vol. 7, pp. 145–187 (2011).

- Heideman, Michael T.; Burrus, C. Sidney (1986). "On the number of multiplications necessary to compute a length-2^n DFT". *IEEE Trans. Acoust. Speech. Sig. Proc.* **34** (1): 91–95. doi:10.1109/TASSP.1986.1164785.

- Johnson, S. G.; Frigo, M. (2007). "A modified split-radix FFT with fewer arithmetic operations" (PDF). *IEEE Trans. Signal Processing* **55** (1): 111–119. doi:10.1109/tsp.2006.882087.

- T. Lundy and J. Van Buskirk, 2007. "A new matrix approach to real FFTs and convolutions of length 2^k," *Computing* **80** (1): 23–45.

- Kent, Ray D. and Read, Charles (2002). *Acoustic Analysis of Speech.* ISBN 0-7693-0112-6. Cites Strang, G. (1994)/May–June). Wavelets. *American Scientist, 82,* 250–255.

- Morgenstern, Jacques (1973). "Note on a lower bound of the linear complexity of the fast Fourier transform". *J. ACM* **20** (2): 305–306. doi:10.1145/321752.321761.

- Mohlenkamp, M. J. (1999). "A fast transform for spherical harmonics" (PDF). *J. Fourier Anal. Appl.* **5** (2–3): 159–184. doi:10.1007/BF01261607.

- Nussbaumer, H. J. (1977). "Digital filtering using polynomial transforms". *Electronics Lett.* **13** (13): 386–387. doi:10.1049/el:19770280.

- V. Pan, 1986, The trade-off between the additive complexity and the asyncronicity of linear and bilinear algorithms, *Information Proc. Lett.* **22**: 11–14.

- Christos H. Papadimitriou, 1979, Optimality of the fast Fourier transform, *J. ACM* **26**: 95–102.

- D. Potts, G. Steidl, and M. Tasche, 2001. "Fast Fourier transforms for nonequispaced data: A tutorial", in: J.J. Benedetto and P. Ferreira (Eds.), *Modern Sampling Theory: Mathematics and Applications* (Birkhauser).

- Press, WH; Teukolsky, SA; Vetterling, WT; Flannery, BP (2007), "Chapter 12. Fast Fourier Transform", *Numerical Recipes: The Art of Scientific Computing* (3rd ed.), New York: Cambridge University Press, ISBN 978-0-521-88068-8

- Rokhlin, Vladimir; Tygert, Mark (2006). "Fast algorithms for spherical harmonic expansions". *SIAM J. Sci. Computing* **27** (6): 1903–1928. doi:10.1137/050623073.

- Schatzman, James C. (1996). "Accuracy of the discrete Fourier transform and the fast Fourier transform". *SIAM J. Sci. Comput* **17**: 1150–1166. doi:10.1137/s1064827593247023.

- Shentov, O. V.; Mitra, S. K.; Heute, U.; Hossen, A. N. (1995). "Subband DFT. I. Definition, interpretations and extensions". *Signal Processing* **41** (3): 261–277. doi:10.1016/0165-1684(94)00103-7.

- Sorensen, H. V.; Jones, D. L.; Heideman, M. T.; Burrus, C. S. (1987). "Real-valued fast Fourier transform algorithms". *IEEE Trans. Acoust. Speech Sig. Processing* **35** (35): 849–863. doi:10.1109/TASSP.1987.1165220. See also Sorensen, H.; Jones, D.; Heideman, M.; Burrus, C. (1987). "Corrections to "Real-valued fast Fourier transform algorithms"".*IEEE Transactions on Acoustics, Speech, and Signal Processing***35**(9): 1353–1353. doi:10.1109/TAS

- Welch, Peter D. (1969). "A fixed-point fast Fourier transform error analysis". *IEEE Trans. Audio Electroacoustics* **17** (2): 151–157. doi:10.1109/TAU.1969.1162035.

- Winograd, S. (1978). "On computing the discrete Fourier transform". *Math. Computation* **32** (141): 175–199. doi:10.1090/S0025-5718-1978-0468306-4. JSTOR 2006266.

8.16   External links

- Fast Fourier Algorithm

- *Fast Fourier Transforms*, Connexions online book edited by C. Sidney Burrus, with chapters by C. Sidney Burrus, Ivan Selesnick, Markus Pueschel, Matteo Frigo, and Steven G. Johnson (2008).

- Links to FFT code and information online.

- National Taiwan University – FFT

- FFT programming in C++ — Cooley–Tukey algorithm.

- Online documentation, links, book, and code.

- Using FFT to construct aggregate probability distributions

- Sri Welaratna, "Thirty years of FFT analyzers", *Sound and Vibration* (January 1997, 30th anniversary issue). A historical review of hardware FFT devices.

- FFT Basics and Case Study Using Multi-Instrument

- FFT Textbook notes, PPTs, Videos at Holistic Numerical Methods Institute.

- ALGLIB FFT Code GPL Licensed multilanguage (VBA, C++, Pascal, etc.) numerical analysis and data processing library.

- MIT's sFFT MIT Sparse FFT algorithm and implementation.

- VB6 FFT VB6 optimized library implementation with source code.

- Fast Fourier transform illustrated Demo examples and FFT calculator.

Chapter 9

Time–frequency analysis

See also: Time–frequency representation

In signal processing, **time–frequency analysis** comprises those techniques that study a signal in both the time and frequency domains *simultaneously,* using various time–frequency representations. Rather than viewing a 1-dimensional signal (a function, real or complex-valued, whose domain is the real line) and some transform (another function whose domain is the real line, obtained from the original via some transform), time–frequency analysis studies a two-dimensional signal – a function whose domain is the two-dimensional real plane, obtained from the signal via a time–frequency transform.[1] [2]

The mathematical motivation for this study is that functions and their transform representation are often tightly connected, and they can be understood better by studying them jointly, as a two-dimensional object, rather than separately. A simple example is that the 4-fold periodicity of the Fourier transform – and the fact that two-fold Fourier transform reverses direction – can be interpreted by considering the Fourier transform as a 90° rotation in the associated time–frequency plane: 4 such rotations yield the identity, and 2 such rotations simply reverse direction (reflection through the origin).

The practical motivation for time–frequency analysis is that classical Fourier analysis assumes that signals are infinite in time or periodic, while many signals in practice are of short duration, and change substantially over their duration. For example, traditional musical instruments do not produce infinite duration sinusoids, but instead begin with an attack, then gradually decay. This is poorly represented by traditional methods, which motivates time–frequency analysis.

One of the most basic forms of time–frequency analysis is the short-time Fourier transform (STFT), but more sophisticated techniques have been developed, notably wavelets.

9.1   Need for a time–frequency approach

In signal processing, **time–frequency analysis** [3] is a body of techniques and methods used for characterizing and manipulating signals whose statistics vary in time, such as transient signals.

It is a generalization and refinement of Fourier analysis, for the case when the signal frequency characteristics are varying with time. Since many signals of interest – such as speech, music, images, and medical signals – have changing frequency characteristics, time–frequency analysis has broad scope of applications.

Whereas the technique of the Fourier transform can be extended to obtain the frequency spectrum of any slowly growing locally integrable signal, this approach requires a complete description of the signal's behavior over all time. Indeed, one can think of points in the (spectral) frequency domain as smearing together information from across the entire time domain. While mathematically elegant, such a technique is not appropriate for analyzing a signal with indeterminate future behavior. For instance, one must presuppose some degree of indeterminate future behavior in any telecommunications systems to achieve non-zero entropy (if one already knows what the other person will say one cannot learn anything).

To harness the power of a frequency representation without the need of a complete characterization in the time domain, one first obtains a time–frequency distribution of the signal, which represents the signal in both the time and frequency domains simultaneously. In such a representation the frequency domain will only reflect the behavior of a temporally localized version of the signal. This enables one to talk sensibly about signals whose component frequencies vary in time.

For instance rather than using tempered distributions to globally transform the following function into the frequency domain one could instead use these methods to describe it as a signal with a time varying frequency.

$$x(t) = \begin{cases} \cos(\pi t); & t < 10 \\ \cos(3\pi t); & 10 \le t < 20 \\ \cos(2\pi t); & t > 20 \end{cases}$$

Once such a representation has been generated other techniques in time–frequency analysis may then be applied to the signal in order to extract information from the signal, to separate the signal from noise or interfering signals, etc.

9.2 Time–frequency distribution functions

9.2.1 Diversity of time–frequency formulations

There are several different ways to formulate a valid time–frequency distribution function, resulting in several well-known time–frequency distributions, such as:

- Short-time Fourier transform (including the Gabor transform),

- Wavelet transform,

- Bilinear time–frequency distribution function (Wigner distribution function, or WDF),

- Modified Wigner distribution function, Gabor–Wigner distribution function, and so on (see Gabor–Wigner transform).

More information about the history and the motivation of development of time–frequency distribution can be found in the entry Time–frequency representation.

9.2.2 Ideal TF distribution function

A time–frequency distribution function ideally has the following properties:

1. **High clarity** to make it easier to be analyzed and interpreted.

2. **No cross-term** to avoid confusing real components from artifacts or noise.

3. **A list of desirable mathematical properties** to ensure such methods benefit real-life application.

4. **Lower computational complexity** to ensure the time needed to represent and process a signal on a time–frequency plane allows real-time implementations.

Below is a brief comparison of some selected time–frequency distribution functions.

To analyze the signals well, choosing an appropriate time–frequency distribution function is important. Which time–frequency distribution function should be used depends on the application being considered, as shown by reviewing a list of applications.[4] The high clarity of the Wigner distribution function (WDF) obtained for some signals is due to the auto-correlation function inherent in its formulation; however, the latter also causes the cross-term problem. Therefore, if we want to analyze a single-term signal, using the WDF may be the best approach; if the signal is composed of multiple components, some other methods like the Gabor transform, Gabor-Wigner distribution or Modified B-Distribution functions may be better choices.

As an illustration, Fourier analysis cannot distinguish the signals:

$$x_1(t) = \begin{cases} \cos(\pi t); & t < 10 \\ \cos(3\pi t); & 10 \le t < 20 \\ \cos(2\pi t); & t > 20 \end{cases}$$

$$x_2(t) = \begin{cases} \cos(\pi t); & t < 10 \\ \cos(2\pi t); & 10 \le t < 20 \\ \cos(3\pi t); & t > 20 \end{cases}$$

But time–frequency analysis can.

9.3 Signal processing applications

The following applications need not only the time–frequency distribution functions but also some operations to the signal. The Linear canonical transform (LCT) is really helpful. By LCTs, the shape and location on the time–frequency plane of a signal can be in the arbitrary form that we want it to be. For example, the LCTs can shift the time–frequency distribution to any location, dilate it in the horizontal and vertical direction without changing its area on the plane, shear (or twist) it, and rotate it (Fractional Fourier transform). This powerful operation, LCT, make it more flexible to analyze and apply the time–frequency distributions.

9.3.1 Instantaneous frequency estimation

The definition of instantaneous frequency is the time rate of change of phase, or

$$\frac{1}{2\pi} \frac{d}{dt} \phi(t),$$

where $\phi(t)$ is the instantaneous phase of a signal. We can know the instantaneous frequency from the time–frequency plane directly if the image is clear enough. Because the high clarity is critical, we often use WDF to analyze it.

9.3.2 TF filtering and signal decomposition

The goal of filter design is to remove the undesired component of a signal. Conventionally, we can just filter in the time domain or in the frequency domain individually as shown below.

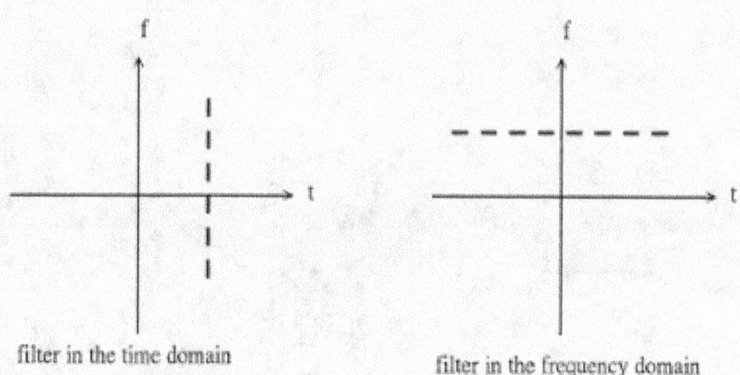

filter in the time domain   filter in the frequency domain

The filtering methods mentioned above can't work well for every signal which may overlap in the time domain or in the frequency domain. By using the time–frequency distribution function, we can filter in the Euclidean time–frequency domain or in the fractional domain by employing the fractional Fourier transform. An example is shown below.

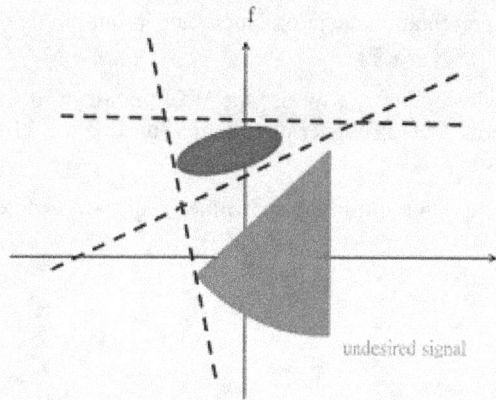

Filter design in time–frequency analysis always deals with signals composed of multiple components, so one cannot use WDF due to cross-term. The Gabor transform, Gabor-Wigner distribution function, or Cohen's class distribution function may be better choices.

The concept of signal decomposition relates to the need to separate one component from the others in a signal; this can be achieved through a filtering operation which require a filter design stage. Such filtering is traditionally done in the time domain or in the frequency domain; however, this may not be possible in the case of non-stationary signals that are multicomponent as such components could overlap in both the time domain and also in the frequency domain; as a consequence, the only possible way to achieve component separation and therefore a signal decomposition is to implement a time–frequency filter.

9.3.3   Sampling theory

By the Nyquist–Shannon sampling theorem, we can conclude that the minimum number of sampling points without aliasing is equivalent to the area of the time–frequency distribution of a signal. (This is actually just an approximation, because the TF area of any signal is infinite.) Below is an example before and after we combine the sampling theory with the time–frequency distribution:

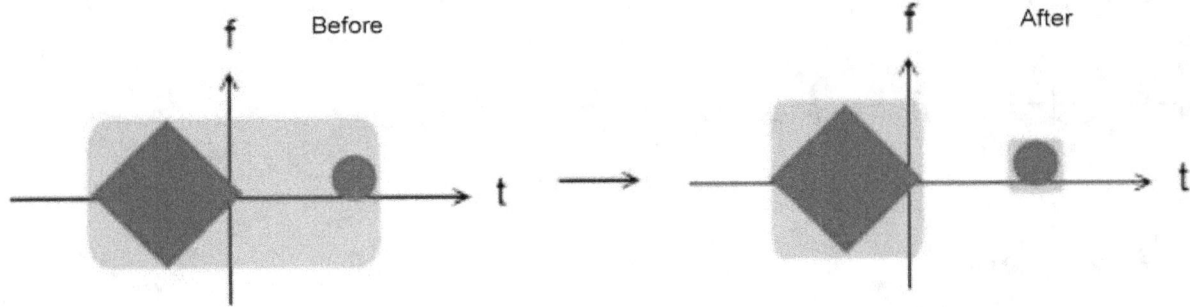

It is noticeable that the number of sampling points decreases after we apply the time–frequency distribution.

When we use the WDF, there might be the cross-term problem (also called interference). On the other hand, using Gabor transform causes an improvement in the clarity and readability of the representation, therefore improving its interpretation and application to practical problems.

Consequently, when the signal we tend to sample is composed of single component, we use the WDF; however, if the signal consists of more than one component, using the Gabor transform, Gabor-Wigner distribution function, or other reduced interference TFDs may achieve better results.

The Balian–Low theorem formalizes this, and provides a bound on the minimum number of time–frequency samples needed.

9.4 Other applications

9.4.1 Modulation and multiplexing

Conventionally, the operation of modulation and multiplexing concentrates in time or in frequency, separately. By taking advantage of the time–frequency distribution, we can make it more efficient to modulate and multiplex. All we have to do is to fill up the time–frequency plane. We present an example as below.

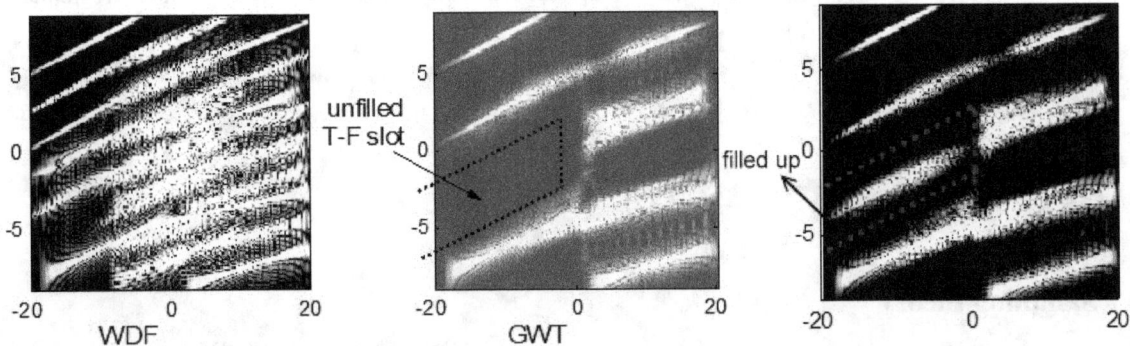

As illustrated in the upper example, using the WDF is not smart since the serious cross-term problem make it difficult to multiplex and modulate.

9.4.2   Electromagnetic wave propagation

We can represent an electromagnetic wave in the form of a 2 by 1 matrix

$$\begin{bmatrix} x \\ y \end{bmatrix},$$

which is similar to the time–frequency plane. When electromagnetic wave propagates through free-space, the Fresnel diffraction occurs. We can operate with the 2 by 1 matrix

$$\begin{bmatrix} x \\ y \end{bmatrix}$$

by LCT with parameter matrix

$$\begin{bmatrix} a & b \\ c & d \end{bmatrix} = \begin{bmatrix} 1 & \lambda z \\ 0 & 1 \end{bmatrix},$$

where z is the propagation distance and λ is the wavelength. When electromagnetic wave pass through a spherical lens or be reflected by a disk, the parameter matrix should be

$$\begin{bmatrix} a & b \\ c & d \end{bmatrix} = \begin{bmatrix} 1 & 0 \\ -\frac{1}{\lambda f} & 1 \end{bmatrix}$$

and

$$\begin{bmatrix} a & b \\ c & d \end{bmatrix} = \begin{bmatrix} 1 & 0 \\ \frac{1}{\lambda R} & 1 \end{bmatrix}$$

respectively, where f is the focal length of the lens and R is the radius of the disk. These corresponding results can be obtained from

$$\begin{bmatrix} a & b \\ c & d \end{bmatrix}\begin{bmatrix} x \\ y \end{bmatrix}.$$

9.4.3   Optics, acoustics, and biomedicine

Light is a kind of electromagnetic wave, so we apply the time–frequency analysis to optics in the same way as to electromagnetic wave propagation. In the same way, a characteristic of acoustic signals is that, often, its frequency varies really severely with time. Because the acoustic signals usually contain a lot of data, it is suitable to use simpler TFDs such as the Gabor transform to analyze the acoustic signals due to the lower computational complexity. If speed is not an issue, then a detailed comparison with well defined criteria should be made before selecting a particular TFD. Another approach is to define a signal dependent TFD that is adapted to the data. In biomedicine, one can use time–frequency distribution to analyze the electromyography (EMG), Electroencephalography (EEG), Electrocardiogram (ECG) or otoacoustic emissions (OAEs).

9.5 History

Early work in time–frequency analysis can be seen in the Haar wavelets (1909) of Alfréd Haar, though these were not significantly applied to signal processing. More substantial work was undertaken by Dennis Gabor, such as Gabor atoms (1947), an early form of wavelets, and the Gabor transform, a modified short-time Fourier transform. The Wigner–Ville distribution (Ville 1948, in a signal processing context) was another foundational step.

Particularly in the 1930s and 1940s, early time–frequency analysis developed in concert with quantum mechanics (Wigner developed the Wigner–Ville distribution in 1932 in quantum mechanics, and Gabor was influenced by quantum mechanics – see Gabor atom); this is reflected in the shared mathematics of the position-momentum plane and the time–frequency plane – as in the Heisenberg uncertainty principle (quantum mechanics) and the Gabor limit (time–frequency analysis), ultimately both reflecting a symplectic structure.

An early practical motivation for time–frequency analysis was the development of radar – see ambiguity function.

9.6 References

[1] L. Cohen, "Time–Frequency Analysis," *Prentice-Hall*, New York, 1995. ISBN 978-0135945322

[2] E. Sejdić, I. Djurović, J. Jiang, "Time-frequency feature representation using energy concentration: An overview of recent advances," Digital Signal Processing, vol. 19, no. 1, pp. 153-183, January 2009.

[3] P. Flandrin, "Time–frequency/Time–Scale Analysis," *Wavelet Analysis and its Applications*, Vol. 10 *Academic Press*, San Diego, 1999.

[4] A. Papandreou-Suppappola, Applications in Time–Frequency Signal Processing (CRC Press, Boca Raton, Fla., 2002)

9.7 See also

- History of wavelets
- Time–frequency analysis for music signal
- Cone-shape distribution function
- Spectral density estimation

Chapter 10

Short-time Fourier transform

The **short-time Fourier transform (STFT)**, or alternatively **short-term Fourier transform**, is a Fourier-related transform used to determine the sinusoidal frequency and phase content of local sections of a signal as it changes over time.[1] In practice, the procedure for computing STFTs is to divide a longer time signal into shorter segments of equal length and then compute the Fourier transform separately on each shorter segment. This reveals the Fourier spectrum on each shorter segment. One then usually plots the changing spectra as a function of time.

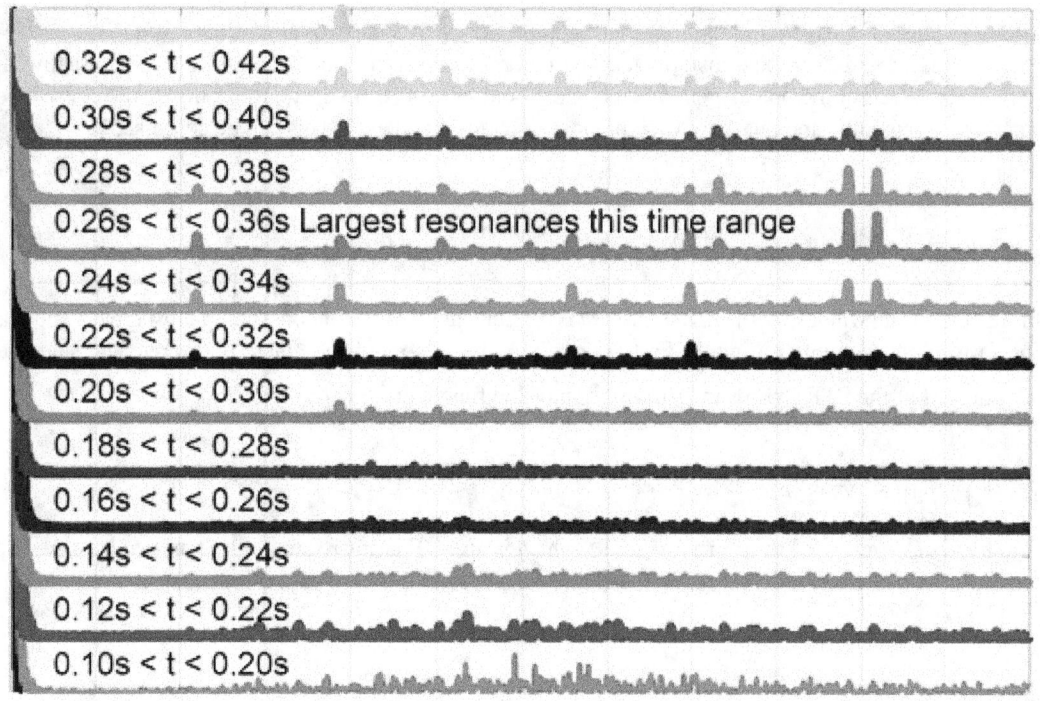

Example of short time Fourier transforms used to determine time of impact from audio signal.

10.1 STFT

10.1.1 Continuous-time STFT

Simply, in the continuous-time case, the function to be transformed is multiplied by a window function which is nonzero for only a short period of time. The Fourier transform (a one-dimensional function) of the resulting signal is taken as the window is slid along the time axis, resulting in a two-dimensional representation of the signal. Mathematically, this is written as:

$$\textbf{STFT}\{x(t)\}(\tau, \omega) \equiv X(\tau, \omega) = \int_{-\infty}^{\infty} x(t)w(t - \tau)e^{-j\omega t}\, dt$$

where $w(t)$ is the window function, commonly a Hann window or Gaussian window centered around zero, and $x(t)$ is the signal to be transformed. (Note the difference between w and ω.) $X(\tau,\omega)$ is essentially the Fourier Transform of $x(t)w(t\text{-}\tau)$, a complex function representing the phase and magnitude of the signal over time and frequency. Often phase unwrapping is employed along either or both the time axis, τ, and frequency axis, ω, to suppress any jump discontinuity of the phase result of the STFT. The time index τ is normally considered to be "*slow*" time and usually not expressed in as high resolution as time t.

10.1.2 Discrete-time STFT

See also: Modified discrete cosine transform

In the discrete time case, the data to be transformed could be broken up into chunks or frames (which usually overlap each other, to reduce artifacts at the boundary). Each chunk is Fourier transformed, and the complex result is added to a matrix, which records magnitude and phase for each point in time and frequency. This can be expressed as:

$$\textbf{STFT}\{x[n]\}(m, \omega) \equiv X(m, \omega) = \sum_{n=-\infty}^{\infty} x[n]w[n - m]e^{-j\omega n}$$

likewise, with signal $x[n]$ and window $w[n]$. In this case, m is discrete and ω is continuous, but in most typical applications the STFT is performed on a computer using the Fast Fourier Transform, so both variables are discrete and quantized.

The magnitude squared of the STFT yields the spectrogram of the function:

$$\text{spectrogram}\{x(t)\}(\tau, \omega) \equiv |X(\tau, \omega)|^2$$

See also the modified discrete cosine transform (MDCT), which is also a Fourier-related transform that uses overlapping windows.

Sliding DFT

If only a small number of ω are desired, or if the STFT is desired to be evaluated for every shift m of the window, then the STFT may be more efficiently evaluated using a sliding DFT algorithm.[2]

10.2 Inverse STFT

The STFT is invertible, that is, the original signal can be recovered from the transform by the Inverse STFT. The most widely accepted way of inverting the STFT is by using the overlap-add (OLA) method, which also allows for modifications to the STFT complex spectrum. This makes for a versatile signal processing method,[3] referred to as the *overlap and add with modifications* method.

10.2.1   Continuous-time STFT

Given the width and definition of the window function $w(t)$, we initially require the area of the window function to be scaled so that

$$\int_{-\infty}^{\infty} w(\tau)\, d\tau = 1.$$

It easily follows that

$$\int_{-\infty}^{\infty} w(t-\tau)\, d\tau = 1 \quad \forall\, t$$

and

$$x(t) = x(t)\int_{-\infty}^{\infty} w(t-\tau)\, d\tau = \int_{-\infty}^{\infty} x(t)w(t-\tau)\, d\tau.$$

The continuous Fourier Transform is

$$X(\omega) = \int_{-\infty}^{\infty} x(t)e^{-j\omega t}\, dt.$$

Substituting $x(t)$ from above:

$$X(\omega) = \int_{-\infty}^{\infty} \left[\int_{-\infty}^{\infty} x(t)w(t-\tau)\, d\tau\right] e^{-j\omega t}\, dt$$

$$= \int_{-\infty}^{\infty}\int_{-\infty}^{\infty} x(t)w(t-\tau)\, e^{-j\omega t}\, d\tau\, dt.$$

Swapping order of integration:

$$X(\omega) = \int_{-\infty}^{\infty}\int_{-\infty}^{\infty} x(t)w(t-\tau)\, e^{-j\omega t}\, dt\, d\tau$$

$$= \int_{-\infty}^{\infty}\left[\int_{-\infty}^{\infty} x(t)w(t-\tau)\, e^{-j\omega t}\, dt\right] d\tau$$

$$= \int_{-\infty}^{\infty} X(\tau,\omega)\, d\tau.$$

So the Fourier Transform can be seen as a sort of phase coherent sum of all of the STFTs of $x(t)$. Since the inverse Fourier transform is

$$x(t) = \frac{1}{2\pi} \int_{-\infty}^{\infty} X(\omega)e^{+j\omega t} \, d\omega,$$

then $x(t)$ can be recovered from $X(\tau,\omega)$ as

$$x(t) = \frac{1}{2\pi} \int_{-\infty}^{\infty} \int_{-\infty}^{\infty} X(\tau,\omega)e^{+j\omega t} \, d\tau \, d\omega.$$

or

$$x(t) = \int_{-\infty}^{\infty} \left[\frac{1}{2\pi} \int_{-\infty}^{\infty} X(\tau,\omega)e^{+j\omega t} \, d\omega \right] d\tau.$$

It can be seen, comparing to above that windowed "grain" or "wavelet" of $x(t)$ is

$$x(t)w(t-\tau) = \frac{1}{2\pi} \int_{-\infty}^{\infty} X(\tau,\omega)e^{+j\omega t} \, d\omega.$$

the inverse Fourier transform of $X(\tau,\omega)$ for τ fixed.

10.3   Resolution issues

Further information: Gabor limit

One of the pitfalls of the STFT is that it has a fixed resolution. The width of the windowing function relates to how the signal is represented—it determines whether there is good frequency resolution (frequency components close together can be separated) or good time resolution (the time at which frequencies change). A wide window gives better frequency resolution but poor time resolution. A narrower window gives good time resolution but poor frequency resolution. These are called narrowband and wideband transforms, respectively.

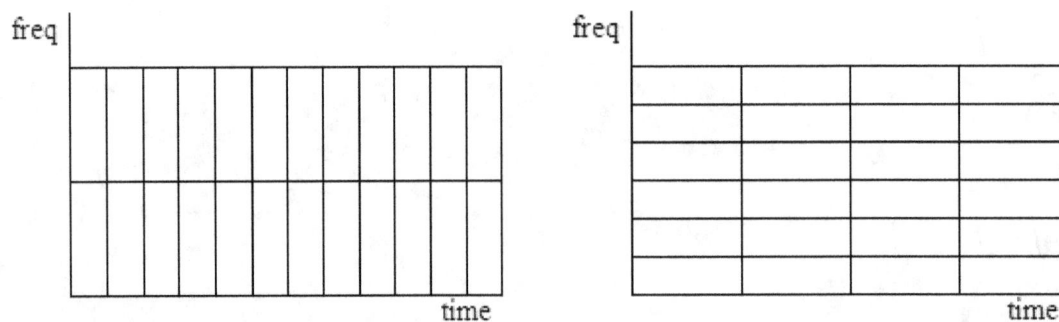

Comparison of STFT resolution. Left has better time resolution, and right has better frequency resolution.

This is one of the reasons for the creation of the wavelet transform and multiresolution analysis, which can give good time resolution for high-frequency events and good frequency resolution for low-frequency events, the combination best suited for many real signals.

This property is related to the Heisenberg uncertainty principle, but not directly – see Gabor limit for discussion. The product of the standard deviation in time and frequency is limited. The boundary of the uncertainty principle (best simultaneous resolution of both) is reached with a Gaussian window function, as the Gaussian minimizes the Fourier uncertainty principle. This is called the Gabor transform (and with modifications for multiresolution becomes the Morlet wavelet transform).

One can consider the STFT for varying window size as a two-dimensional domain (time and frequency), as illustrated in the example below, which can be calculated by varying the window size. However, this is no longer a strictly time–frequency representation – the kernel is not constant over the entire signal.

10.3.1   Example

Using the following sample signal $x(t)$ that is composed of a set of four sinusoidal waveforms joined together in sequence. Each waveform is only composed of one of four frequencies (10, 25, 50, 100 Hz). The definition of $x(t)$ is:

$$x(t) = \begin{cases} \cos(2\pi 10t) & 0\,\mathrm{s} \le t < 5\,\mathrm{s} \\ \cos(2\pi 25t) & 5\,\mathrm{s} \le t < 10\,\mathrm{s} \\ \cos(2\pi 50t) & 10\,\mathrm{s} \le t < 15\,\mathrm{s} \\ \cos(2\pi 100t) & 15\,\mathrm{s} \le t < 20\,\mathrm{s} \end{cases}$$

Then it is sampled at 400 Hz. The following spectrograms were produced:

The 25 ms window allows us to identify a precise time at which the signals change but the precise frequencies are difficult to identify. At the other end of the scale, the 1000 ms window allows the frequencies to be precisely seen but the time between frequency changes is blurred.

10.3.2   Explanation

It can also be explained with reference to the sampling and Nyquist frequency.

Take a window of N samples from an arbitrary real-valued signal at sampling rate f_s. Taking the Fourier transform produces N complex coefficients. Of these coefficients only half are useful (the last $N/2$ being the complex conjugate of the first $N/2$ in reverse order, as this is a real valued signal).

These $N/2$ coefficients represent the frequencies 0 to $f_\mathrm{s}/2$ (Nyquist) and two consecutive coefficients are spaced apart by f_s/N Hz.

To increase the frequency resolution of the window the frequency spacing of the coefficients needs to be reduced. There are only two variables, but decreasing f_s (and keeping N constant) will cause the window size to increase — since there are now fewer samples per unit time. The other alternative is to increase N, but this again causes the window size to increase. So any attempt to increase the frequency resolution causes a larger window size and therefore a reduction in time resolution—and vice versa.

10.4   Rayleigh frequency

As the Nyquist frequency is a limitation in the maximum frequency that can be meaningfully analysed, so is the Rayleigh frequency a limitation on the minimum frequency.

Rayleigh frequency is the minimum frequency that can be resolved by a finite duration time window.[4][5]

Given a time window that is T seconds long, the minimum frequency that can be resolved is 1/T Hz.

Rayleigh frequency is important to consider in applications of the short-time Fourier transform (STFT), such as in analysing neural signals[6][7]

10.5 Application

An STFT being used to analyze an audio signal across time

STFTs as well as standard Fourier transforms and other tools are frequently used to analyze music. The spectrogram can, for example, show frequency on the horizontal axis, with the lowest frequencies at left, and the highest at the right. The height of each bar (augmented by color) represents the amplitude of the frequencies within that band. The depth dimension represents time, where each new bar was a separate distinct transform. Audio engineers use this kind of visual to gain information about an audio sample, for example, to locate the frequencies of specific noises (especially when used with greater frequency resolution) or to find frequencies which may be more or less resonant in the space where the signal was recorded. This information can be used for equalization or tuning other audio effects.

10.6 See also

- Spectral density estimation
- Time-frequency representation
- Reassignment method

Other time-frequency transforms:

- wavelet transform

- chirplet transform

- fractional Fourier transform

- Newland transform

- Constant Q transform

- Gabor transform

- cone-shape distribution function

10.7   References

[1] E. Sejdić, I. Djurović, J. Jiang, "Time-frequency feature representation using energy concentration: An overview of recent advances," Digital Signal Processing, vol. 19, no. 1, pp. 153-183, January 2009.

[2] E. Jacobsen and R. Lyons, The sliding DFT, *Signal Processing Magazine* vol. 20, issue 2, pp. 74–80 (March 2003).

[3] Jont B. Allen (June 1977). "Short Time Spectral Analysis, Synthesis, and Modification by Discrete Fourier Transform". *IEEE Transactions on Acoustics, Speech, and Signal Processing.* ASSP-25 (3): 235–238.

[4] https://physics.ucsd.edu/neurophysics/publications/Cold%20Spring%20Harb%20Protoc-2014-Kleinfeld-pdb.top081075.pdf

[5] http://fieldtrip.fcdonders.nl/faq/what_does_padding_not_sufficient_for_requested_frequency_resolution_mean

[6] http://www.ncbi.nlm.nih.gov/pmc/articles/PMC2441488

[7] http://www.jneurosci.org/content/30/20/7078.full

10.8   External links

- DiscreteTFDs – software for computing the short-time Fourier transform and other time-frequency distributions

- Singular Spectral Analysis - MultiTaper Method Toolkit - a free software program to analyze short, noisy time series.

- kSpectra Toolkit for Mac OS X from SpectraWorks

- Time stretched short time Fourier transform for time frequency analysis of ultra wideband signals

- A BSD-licensed Matlab class to perform STFT and inverse STFT

- LTFAT - A free (GPL) Matlab / Octave toolbox to work with short-time Fourier transforms and time-frequency analysis

Chapter 11

Gabor transform

The **Gabor transform**, named after Dennis Gabor, is a special case of the short-time Fourier transform. It is used to determine the sinusoidal frequency and phase content of local sections of a signal as it changes over time. The function to be transformed is first multiplied by a Gaussian function, which can be regarded as a window function, and the resulting function is then transformed with a Fourier transform to derive the time-frequency analysis.[1] The window function means that the signal near the time being analyzed will have higher weight. The Gabor transform of a signal x(t) is defined by this formula:

$$G_x(t, f) = \int_{-\infty}^{\infty} e^{-\pi(\tau - t)^2} e^{-j2\pi f\tau} x(\tau)\, d\tau$$

The Gaussian function has infinite range and it is impractical for implementation. However, a level of significance can be chosen (for instance 0.00001) for the distribution of the Gaussian function.

$$\begin{cases} e^{-\pi a^2} \geq 0.00001; & |a| \leq 1.9143 \\ e^{-\pi a^2} < 0.00001; & |a| > 1.9143 \end{cases}$$

Outside these limits of integration ($|a| > 1.9143$) the Gaussian function is small enough to be ignored. Thus the Gabor transform can be satisfactorily approximated as

$$G_x(t, f) = \int_{-1.9143+t}^{1.9143+t} e^{-\pi(\tau - t)^2} e^{-j2\pi f\tau} x(\tau)\, d\tau$$

This simplification makes the Gabor transform practical and realizable.

The window function width can also be varied to optimize the time-frequency resolution tradeoff for a particular application by replacing the $-\pi(\tau - t)^2$ with $-\pi\alpha(\tau - t)^2$ for some chosen alpha.

11.1   Inverse Gabor transform

The Gabor transform is invertible. The original signal can be recovered by the following equation

$$x(t) = \int_{-\infty}^{\infty} \int_{-\infty}^{\infty} G_x(\tau, f) e^{j2\pi tf}\, df\, d\tau$$

Compare this inversion formula with property No. 5 below.

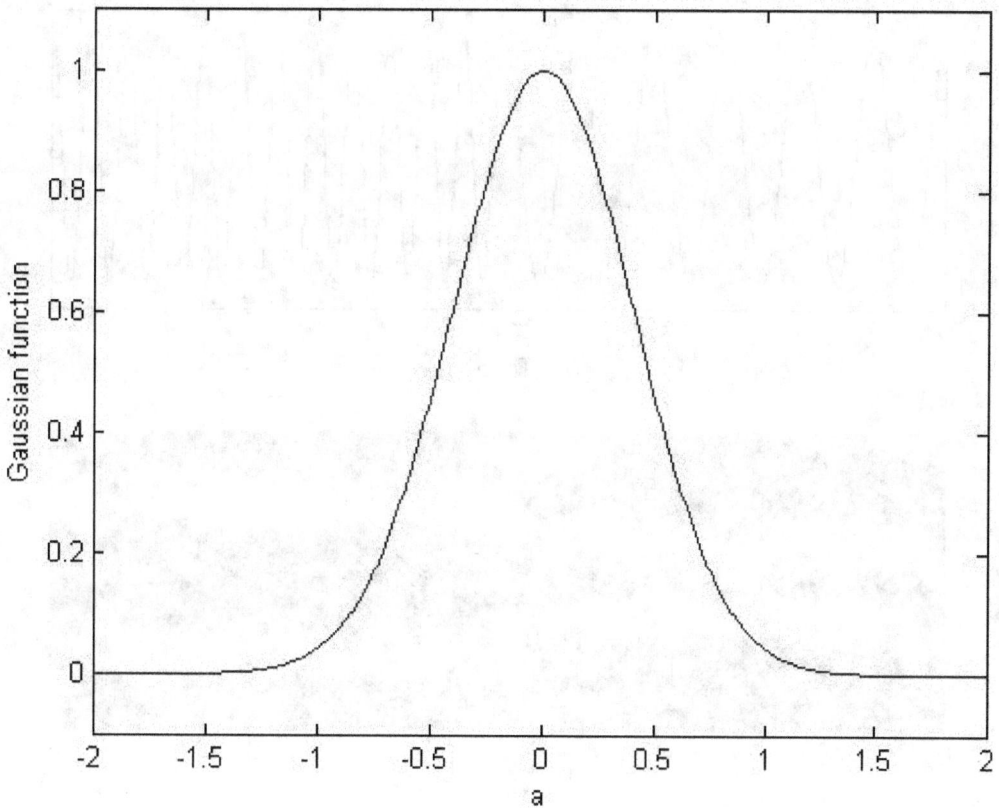

Magnitude of Gaussian function.

11.2   Properties of the Gabor transform

The Gabor transform has many properties like those of the Fourier transform. These properties are listed in the following tables.

11.3   Application and example

The main application of the Gabor transform is used in time frequency analysis. Take the following equation as an example. The input signal has 1 Hz frequency component when $t \leq 0$ and has 2 Hz frequency component when $t > 0$

$$x(t) = \begin{cases} \cos(2\pi t) & \text{for} t \leq 0, \\ \cos(4\pi t) & \text{for} t > 0. \end{cases}$$

But if the total bandwidth available is 5 Hz, other frequency bands except $x(t)$ are wasted. Through time frequency analysis by applying the Gabor transform, the available bandwidth can be known and those frequency bands can be used for other applications and bandwidth is saved. The right side picture show the input signal $x(t)$ and the output of the Gabor transform. As was our expectation, the frequency distribution can be separated into two parts. One is $t \leq 0$ and the other is $t > 0$. The white part is the frequency band occupied by $x(t)$ and the black part is not used.

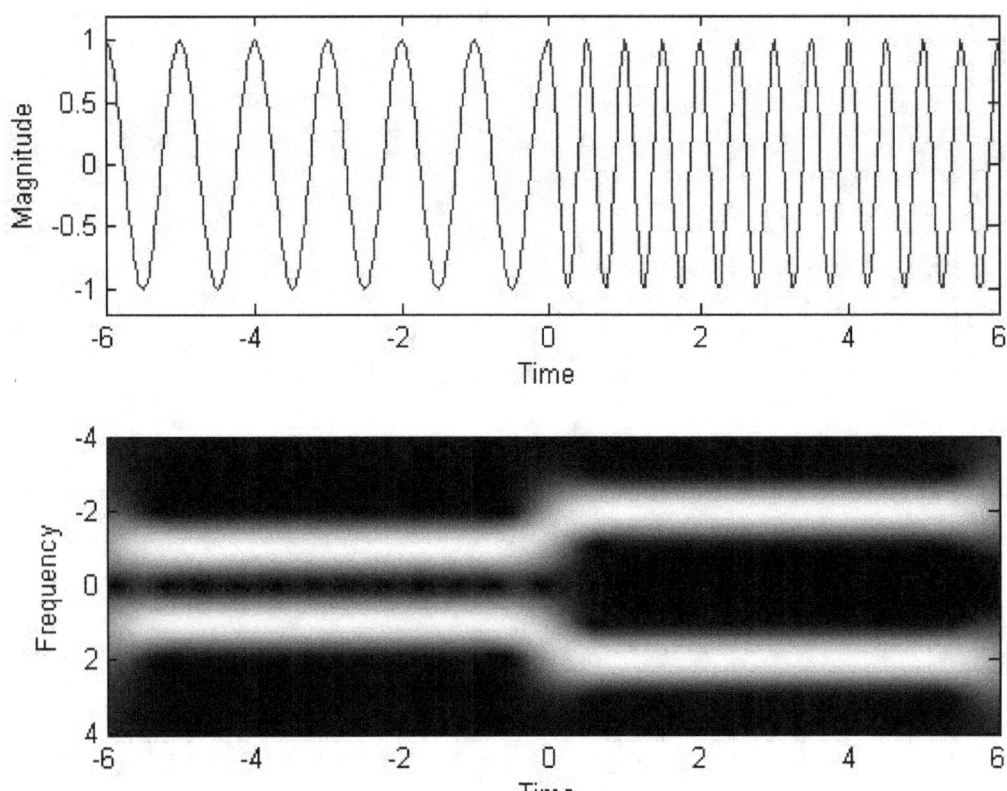

Time/frequency distribution.

11.4   Discrete Gabor-transformation

A discrete version of Gabor representation

$$y(t) = \sum_{m=-\infty}^{\infty} \sum_{n=-\infty}^{\infty} C_{nm} \cdot g_{nm}(t)$$

with $g_{nm}(t) = s(t - m\tau_0) \cdot e^{j\Omega nt}$

can be derived easily by discretizing the Gabor-basis-function in these equations. Hereby the continuous parameter t is replaced by the discrete time k. Furthermore the now finite summation limit in Gabor representation has to be considered. In this way, the sampled signal y(k) is split into M time frames of length N. According to $\Omega \leq \frac{2\pi}{\tau_0}$, the factor Ω for critical sampling is $\Omega = \frac{2\pi}{N}$

Similar to the DFT (discrete Fourier transformation) a frequency domain split into N discrete partitions is obtained. An inverse transformation of these N spectral partitions then leads to N values y(k)for the time window, which consists of N sample values. For overall M time windows with N sample values, each signal y(k) contains K=N · M sample values: (the discrete Gabor representation)

$$y(k) = \sum_{m=0}^{M-1} \sum_{n=0}^{N-1} C_{nm} \cdot g_{nm}(k)$$

with $g_{nm}(k) = s(k - mN) \cdot e^{j\Omega nk}$

According to the equation above, the N $\cdot$ M coefficients C_{nm} correspond to the number of sample values K of the signal.

For over-sampling Ω is set to $\Omega \leq \frac{2\pi}{N} = \frac{2\pi}{N'}$ with N' > N, which results in N' > N summation coefficients in the second sum of the discrete Gabor representation. In this case, the number of obtained Gabor-coefficients would be M $\cdot$ N'>K. Hence, more coefficients than sample values are available and therefore a redundant representation would be achieved.

11.5   See also

- Gabor filter

- Gabor wavelet

- Gabor atom

- Time-frequency representation

- S transform

- Short-time Fourier transform

- Wigner distribution function

11.6   References

[1] E. Sejdić, I. Djurović, J. Jiang, "Time-frequency feature representation using energy concentration: An overview of recent advances," *Digital Signal Processing*, vol. 19, no. 1, pp. 153-183, January 2009.

- Jian-Jiun Ding, Time frequency analysis and wavelet transform class note, the Department of Electrical Engineering, National Taiwan University (NTU), Taipei, Taiwan, 2007.

Chapter 12

Fractional Fourier transform

In mathematics, in the area of harmonic analysis, the **fractional Fourier transform** (**FRFT**) is a family of linear transformations generalizing the Fourier transform. It can be thought of as the Fourier transform to the n-th power, where n need not be an integer — thus, it can transform a function to any *intermediate* domain between time and frequency. Its applications range from filter design and signal analysis to phase retrieval and pattern recognition.

The FRFT can be used to define fractional convolution, correlation, and other operations, and can also be further generalized into the linear canonical transformation (LCT). An early definition of the FRFT was introduced by Condon,[1] by solving for the Green's function for phase-space rotations, and also by Namias,[2] generalizing work of Wiener[3] on Hermite polynomials.

However, it was not widely recognized in signal processing until it was independently reintroduced around 1993 by several groups.[4] Since then, there has been a surge of interest in extending Shannon's sampling theorem[5][6] for signals which are band-limited in the Fractional Fourier domain.

A completely different meaning for "fractional Fourier transform" was introduced by Bailey and Swartztrauber[7] as essentially another name for a z-transform, and in particular for the case that corresponds to a discrete Fourier transform shifted by a fractional amount in frequency space (multiplying the input by a linear chirp) and evaluating at a fractional set of frequency points (e.g. considering only a small portion of the spectrum). (Such transforms can be evaluated efficiently by Bluestein's FFT algorithm.) This terminology has fallen out of use in most of the technical literature, however, in preference to the FRFT. The remainder of this article describes the FRFT.

12.1 Introduction

The continuous Fourier transform $\mathcal{F}$ of a function $\mathfrak{f}\colon \mathbf{R} \to \mathbf{C}$ is a unitary operator of L^2 that maps the function $\mathfrak{f}$ to its frequential version $\mathfrak{f}$:

$$\hat{f}(\xi) = \int_{-\infty}^{\infty} f(x)\, e^{-2\pi i x \xi} \, \mathrm{d}x \,, \text{ for every real number } \xi \,.$$

And $\mathfrak{f}$ is determined by $\mathfrak{f}$ via the inverse transform $\mathcal{F}^{-1}$

$$f(x) = \int_{-\infty}^{\infty} \hat{f}(\xi)\, e^{2\pi i \xi x} \, \mathrm{d}\xi, \text{ for every real number } x.$$

Let us study its n-th iterated $\mathcal{F}^n$ defined by $\mathcal{F}^n[f] = \mathcal{F}[\mathcal{F}^{n-1}[f]]$ and $\mathcal{F}^{-n} = (\mathcal{F}^{-1})^n$ when n is a non-negative integer, and $\mathcal{F}^0[f] = f$. Their sequence is finite since $\mathcal{F}$ is a 4-periodic automorphism: for every function $\mathfrak{f}$, $\mathcal{F}^4[f] = f$.

More precisely, let us introduce the **parity operator** $\mathcal{P}$ that inverts time, $\mathcal{P}[f]\colon t \mapsto f(-t)$. Then the following properties hold:

$$\mathcal{F}^0 = \mathrm{Id}, \qquad \mathcal{F}^1 = \mathcal{F}, \qquad \mathcal{F}^2 = \mathcal{P}, \qquad \mathcal{F}^4 = \mathrm{Id}$$

$$\mathcal{F}^3 = \mathcal{F}^{-1} = \mathcal{P} \circ \mathcal{F} = \mathcal{F} \circ \mathcal{P}.$$

The FrFT provides a family of linear transforms that further extends this definition to handle non-integer powers $n = 2\alpha/\pi$ of the FT.

12.2 Definition

For any real α, the α-angle fractional Fourier transform of a function f is denoted by $\mathcal{F}_\alpha(u)$ and defined by

(the square root is defined such that the argument of result lies in the interval $[-\pi/2, \pi/2]$)

If α is an integer multiple of π, then the cotangent and cosecant functions above diverge. However, this can be handled by taking the limit, and leads to a Dirac delta function in the integrand. More directly, since $\mathcal{F}^2(f) = f(-t)$, $\mathcal{F}_\alpha(f)$ must be simply $f(t)$ or $f(-t)$ for α an even or odd multiple of π, respectively.

For $\alpha = \pi/2$, this becomes precisely the definition of the continuous Fourier transform, and for $\alpha = -\pi/2$ it is the definition of the inverse continuous Fourier transform.

The FrFT argument u is neither a spatial one x nor a frequency ξ. We will see why it can be interpreted as linear combination of both coordinates (x,ξ). When we want to distinguish the α-angular fractional domain, we will let x_a denote the argument of $\mathcal{F}_\alpha$.

Remark: with the angular frequency ω convention instead of the frequency one, the FrFT formula is the Mehler kernel,

$$\mathcal{F}_\alpha(f)(\omega) = \sqrt{\frac{1 - i\cot(\alpha)}{2\pi}} e^{i\cot(\alpha)\omega^2/2} \int_{-\infty}^{\infty} e^{-i\csc(\alpha)\omega t + i\cot(\alpha)t^2/2} f(t)\, dt \ .$$

12.2.1 Properties

The operator $\mathcal{F}_\alpha$ has the properties :

- **coherence**

  With the FT powers: if $\alpha \equiv k\pi/2\,[2\pi]$, where k is an integer, then
  $$\mathcal{F}_\alpha = \mathcal{F}^k$$

- **additivity**

  For any real angles $\alpha, \beta,$
  $$\mathcal{F}_{\alpha+\beta} = \mathcal{F}_\alpha \circ \mathcal{F}_\beta = \mathcal{F}_\beta \circ \mathcal{F}_\alpha.$$

- **linearity**

  Let $\mathcal{F}_\alpha$ denote the α-th order fractional transform operator, then
  $$\mathcal{F}_\alpha[\textstyle\sum_k b_k f_k(u)] = \sum_k b_k \mathcal{F}_\alpha[f_k(u)]$$

- **integer order**

   When α is equal to an integer multiple of $\pi/2$, the α-th order fractional Fourier transform is equivalent to the k-th integer power of the ordinary Fourier transform, defined by repeated application. This means that
   $$\mathcal{F}_\alpha = \mathcal{F}_{k\pi/2} = \mathcal{F}^k = (\mathcal{F})^k$$
   Moreover, it has following relation
   $$\mathcal{F}^2 = \mathcal{P} \text{ (parity operator)}$$
   $$\mathcal{F}^3 = \mathcal{F}^{-1} = (\mathcal{F})^{-1} \text{ (inverse transform operator)}$$
   $$\mathcal{F}^4 = \mathcal{F}^0 = \mathcal{I} \text{ (identity operator)}$$
   $$\mathcal{F}^j = \mathcal{F}^{j \mod 4}$$

- **inverse**

$$(\mathcal{F}_\alpha)^{-1} = \mathcal{F}_{-\alpha}$$

- **commutativity**

$$\mathcal{F}_{\alpha_1}\mathcal{F}_{\alpha_2} = \mathcal{F}_{\alpha_2}\mathcal{F}_{\alpha_1}$$

- **Associativity**

$$(\mathcal{F}_{\alpha_1}\mathcal{F}_{\alpha_2})\mathcal{F}_{\alpha_3} = \mathcal{F}_{\alpha_1}(\mathcal{F}_{\alpha_2}\mathcal{F}_{\alpha_3})$$

- **Parseval**

   $$\int f^*(u)g(u)du = \int f_\alpha^*(u)g_\alpha(u)du$$
   This property is similar to unitarity. Energy or norm conservation is a special case.

- **Time reversal**

   Let $\mathcal{P}$ denotes the parity operator. $\mathcal{P}[f(u)] = f(-u)$, then
   $$\mathcal{F}_\alpha\mathcal{P} = \mathcal{P}\mathcal{F}_\alpha$$
   $$\mathcal{F}_\alpha[f(-u)] = f_\alpha(-u)$$

- **Transform of a shifted function**

   Let $\mathcal{SH}(u_0)$ and $\mathcal{PH}(v_0)$ denotes the shift and the phase shift operators, respectively. The definition of $\mathcal{SH}(u_0)$ and $\mathcal{PH}(v_0)$ are as following.
   $$\mathcal{SH}(u_0)[f(u)] = f(u + u_0)$$
   $$\mathcal{PH}(v_0)[f(u)] = e^{j2\pi v_0 u}f(u)$$
   Then
   $$\mathcal{F}_\alpha\mathcal{SH}(u_0) = e^{j\pi u_0^2 \sin\alpha\cos\alpha}\mathcal{PH}(u_0\sin\alpha)\mathcal{SH}(u_0\cos\alpha)\mathcal{F}_\alpha$$
   $$\mathcal{F}_\alpha[f(u + u_0)] = e^{j\pi u_0^2 \sin\alpha\cos\alpha}e^{j2\pi u u_0 \sin\alpha}f_\alpha(u + u_0\cos\alpha)$$

See also: Generalizations of Pauli matrices § Construction: The clock and shift matrices

- **Transform of a scaled function**

Let $M(M)$ and $Q(q)$ denotes the scaling and chirp multiplication operators, respectively. The definition of $M(M)$ and $Q(q)$ are as following.

$$M(M)[f(u)] = |M|^{-1/2} f(u/M)$$

$$Q(q)[f(u)] = e^{-j\pi qu^2} f(u)$$

Then,

$$\mathcal{F}_\alpha M(M) = Q(-\cot(\tfrac{1-\cos^2 \alpha'}{\cos^2 \alpha}\alpha)) \times M(\tfrac{\sin\alpha}{M\sin\alpha'})\mathcal{F}_{\alpha'}$$

$$\mathcal{F}_\alpha[|M|^{-1/2} f(u/M)] = \sqrt{\tfrac{1-j\cot\alpha}{1-jM^2\cot\alpha}} e^{j\pi u^2 \cot(\tfrac{1-\cos^2\alpha'}{\cos^2\alpha}\alpha)} \times f_a(\tfrac{Mu\sin\alpha'}{\sin\alpha})$$

Notice that the fractional Fourier transform of $f(u/M)$ cannot be expressed as a scaled version of $f_\alpha(u)$. Rather, the fractional Fourier transform of $f(u/M)$ turns out to be a scaled and chirp modulated version of $f'_\alpha(u)$ where $\alpha \neq \alpha'$ is a different order

12.2.2   Fractional kernel

The FrFT is an integral transform

$$\mathcal{F}_\alpha f(u) = \int K_\alpha(u,x) f(x)\, dx$$

where the α-angle kernel is

$$K_\alpha(u,x) = \begin{cases} \sqrt{1-i\cot(\alpha)} \exp\left(i\pi(\cot(\alpha)(x^2+u^2) - 2\csc(\alpha)ux)\right) & \text{if } \alpha \text{ is not a multiple of } \pi, \\ \delta(u-x) & \text{if } \alpha \text{ is a multiple of } 2\pi, \\ \delta(u+x) & \text{if } \alpha + \pi \text{ is a multiple of } 2\pi, \end{cases}$$

(the square root is defined such that the argument of result lies in the interval $[-\pi/2, \pi/2]$).

Here again the special cases are consistent with the limit behavior when α approaches a multiple of π.

The FrFT has the same properties as its kernels :

- symmetry: $K_\alpha(u,u') = K_\alpha(u',u)$

- inverse: $K_\alpha^{-1}(u,u') = K_\alpha^*(u,u') = K_{-\alpha}(u',u)$

- additivity: $K_{\alpha+\beta}(u,u') = \int K_\alpha(u,u'')K_\beta(u'',u')\, du''$.

12.2.3   Related transforms

There also exist related fractional generalizations of similar transforms such as the discrete Fourier transform. The **discrete fractional Fourier transform** is defined by Zeev Zalevsky in (Candan, Kutay & Ozaktas 2000) and (Ozaktas, Zalevsky & Kutay 2001, Chapter 6).

Fractional wavelet transform (FRWT):[8] A generalization of the classical wavelet transform (WT) in the fractional Fourier transform (FRFT) domains. The FRWT is proposed in order to rectify the limitations of the WT and the FRFT. This transform not only inherits the advantages of multiresolution analysis of the WT, but also has the capability of signal representations in the fractional domain which is similar to the FRFT. Compared with the existing FRWT, the FRWT (defined by Shi, Zhang, and Liu 2012) can offer signal representations in the time-fractional-frequency plane.

See also the chirplet transform for a related generalization of the Fourier transform.

12.2.4 Generalization

The Fourier transform is essentially bosonic; it works because it is consistent with the superposition principle and related interference patterns. There is also a fermionic Fourier transform.[9] These have been generalized into a supersymmetric FRFT, and a supersymmetric Radon transform.[9] There is also a fractional Radon transform, a symplectic FRFT, and a symplectic wavelet transform.[10] Because quantum circuits are based on unitary operations, they are useful for computing integral transforms as the latter are unitary operators on a function space. A quantum circuit has been designed which implements the FRFT.[11]

12.3 Interpretation of the fractional Fourier transform

Further information: Linear canonical transformation
The usual interpretation of the Fourier transform is as a transformation of a time domain signal into a frequency domain

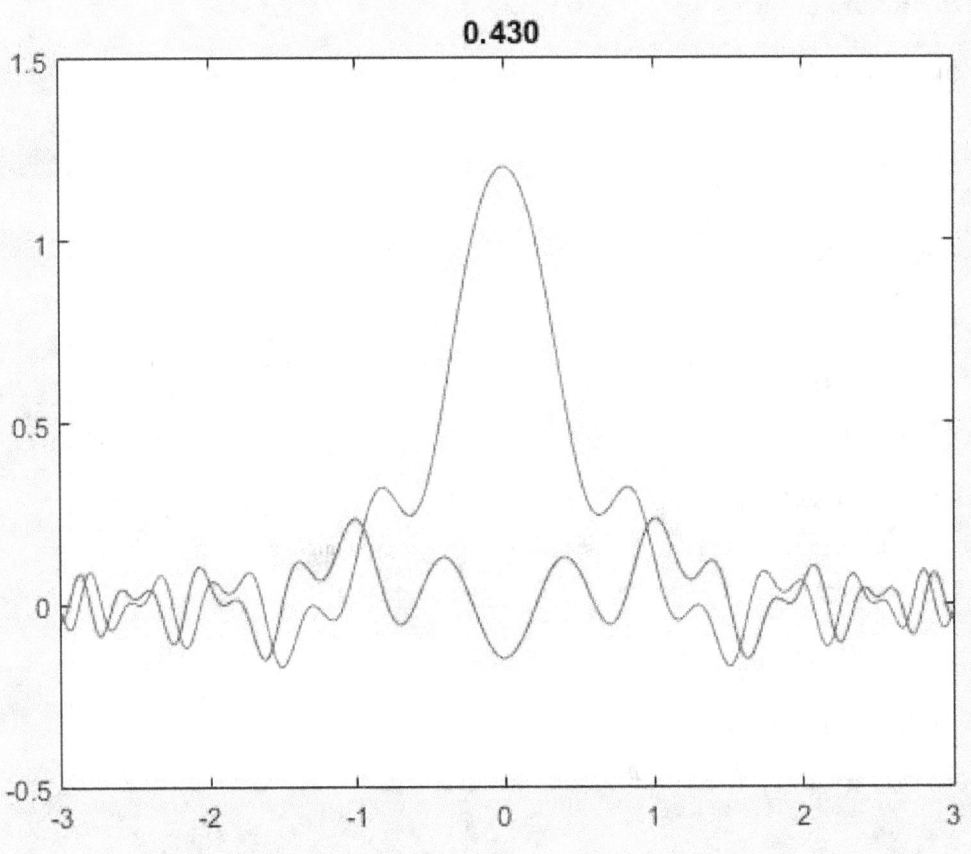

A rect function turns into a sinc function as the order of the Fractional Fourier Transform becomes 1.

signal. On the other hand, the interpretation of the inverse Fourier transform is as a transformation of a frequency domain signal into a time domain signal. Apparently, fractional Fourier transforms can transform a signal (either in the time domain or frequency domain) into the domain between time and frequency: it is a rotation in the time-frequency domain. This perspective is generalized by the linear canonical transformation, which generalizes the fractional Fourier transform and allows linear transforms of the time-frequency domain other than rotation.

Take the below figure as an example. If the signal in the time domain is rectangular (as below), it will become a sinc function in the frequency domain. But if we apply the fractional Fourier transform to the rectangular signal, the transformation output will be in the domain between time and frequency.

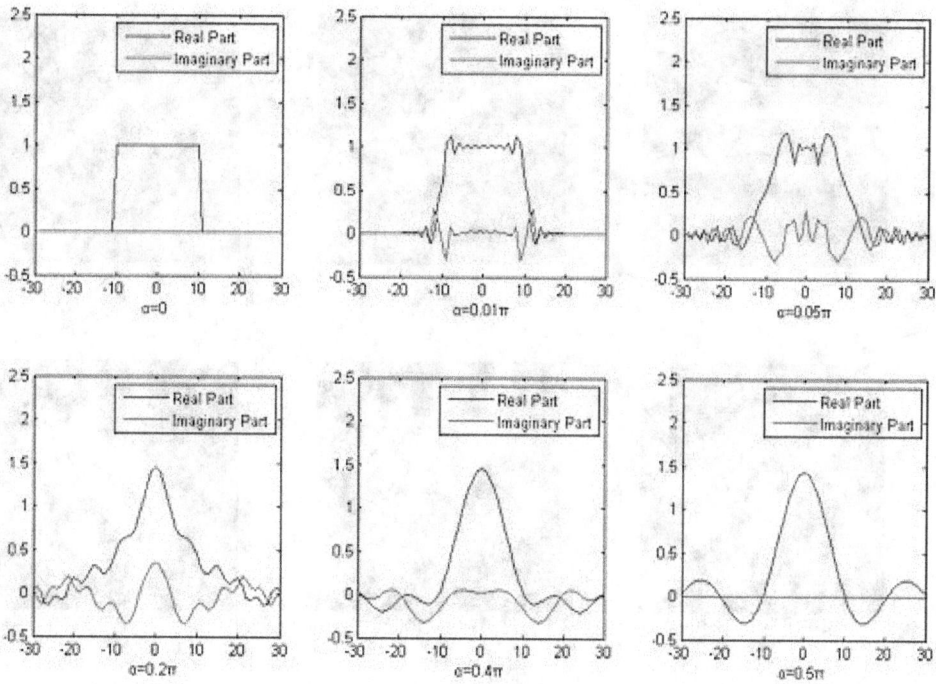

Fractional Fourier transform

Actually, fractional Fourier transform is a rotation operation on the time frequency distribution. From the definition above, for $\alpha = 0$, there will be no change after applying fractional Fourier transform, and for $\alpha = \pi/2$, fractional Fourier transform becomes a Fourier transform, which rotates the time frequency distribution with $\pi/2$. For other value of α, fractional Fourier transform rotates the time frequency distribution according to α. The following figure shows the results of the fractional Fourier transform with different values of α.

12.4   Application

Fractional Fourier transform can be used in time frequency analysis and DSP.[12] It is useful to filter noise, but with the condition that it does not overlap with the desired signal in the time frequency domain. Consider the following example. We cannot apply a filter directly to eliminate the noise, but with the help of the fractional Fourier transform, we can rotate the signal (including the desired signal and noise) first. We then apply a specific filter which will allow only the desired signal to pass. Thus the noise will be removed completely. Then we use the fractional Fourier transform again to rotate the signal back and we can get the desired signal.

Fractional Fourier transforms are also used to design optical systems and optimize holographic storage efficiency.[13]

Thus, using just truncation in the time domain, or equivalently low-pass filters in the frequency domain, one can cut out any convex set in time-frequency space; just using time domain or frequency domain methods without fractional Fourier transforms only allow cutting out rectangles parallel to the axes.

12.5   See also

• Mehler kernel

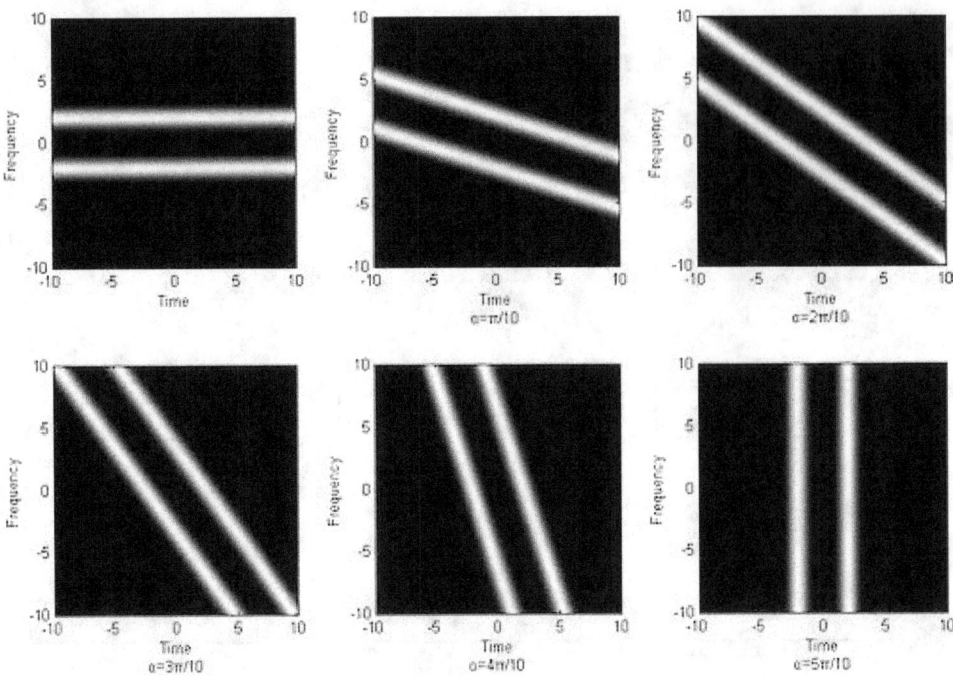

Time/frequency distribution of fractional Fourier transform.

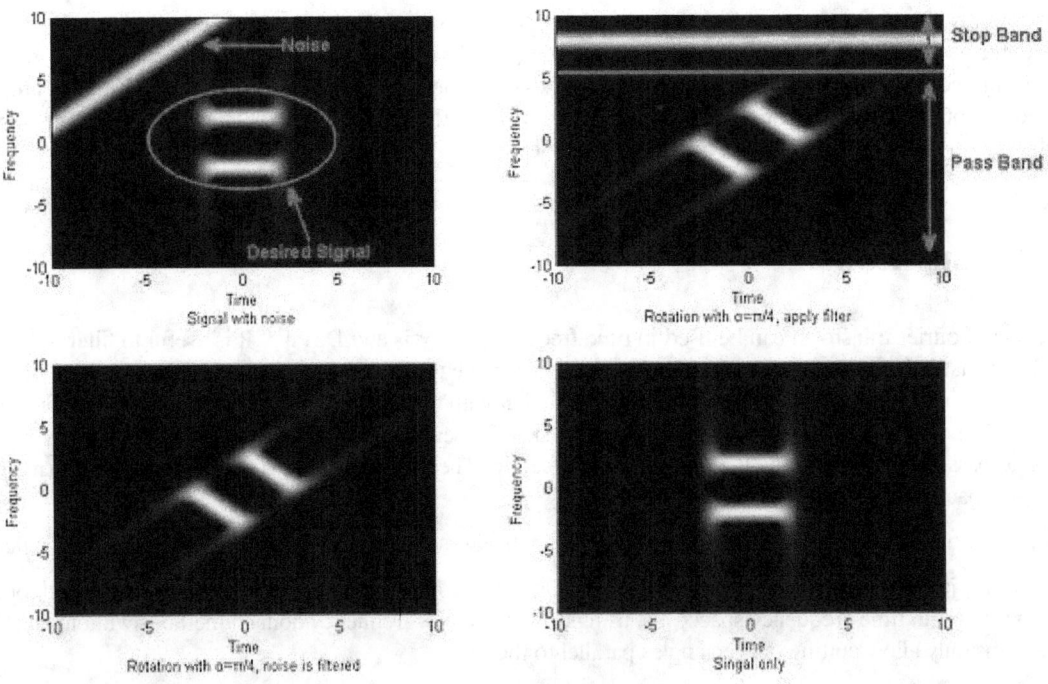

Fractional Fourier transform in DSP.

Other time-frequency transforms:

- Linear canonical transformation

- short-time Fourier transform

- wavelet transform

- chirplet transform

- cone-shape distribution function

12.6   References

[1] E. U. Condon, "Immersion of the Fourier transform in a continuous group of functional transformations", *Proc. Nat. Acad. Sci. USA* **23**, (1937) 158–164. online

[2] V. Namias, "The fractional order Fourier transform and its application to quantum mechanics," *J. Inst. Appl. Math.* **25**, 241–265 (1980).

[3] N. Wiener, "Hermitian Polynomials and Fourier Analysis", *J. Mathematics and Physics* **8** (1929) 70-73.

[4] Luís B. Almeida, "The fractional Fourier transform and time-frequency representations," *IEEE Trans. Sig. Processing* **42** (11), 3084–3091 (1994).

[5] Ran Tao, Bing Deng, Wei-Qiang Zhang and Yue Wang, "Sampling and sampling rate conversion of band limited signals in the fractional Fourier transform domain," *IEEE Transactions on Signal Processing*, **56** (1), 158–171 (2008).

[6] A. Bhandari and P. Marziliano, "Sampling and reconstruction of sparse signals in fractional Fourier domain," *IEEE Signal Processing Letters*, **17** (3), 221–224 (2010).

[7] D. H. Bailey and P. N. Swarztrauber, "The fractional Fourier transform and applications," *SIAM Review* **33**, 389-404 (1991). (Note that this article refers to the chirp-z transform variant, not the FRFT.)

[8] J. Shi, N.-T. Zhang, and X.-P. Liu, "A novel fractional wavelet transform and its applications," Sci. China Inf. Sci. vol. 55, no. 6, pp. 1270-1279, June 2012. URL: http://www.springerlink.com/content/q01np2848m388647/

[9] Hendrik De Bie, *Fourier transform and related integral transforms in superspace (2008)*, http://www.arxiv.org/abs/0805.1918

[10] Hong-yi Fan and Li-yun Hu, *Optical transformation from chirplet to fractional Fourier transformation kernel (2009)*, http://www.arxiv.org/abs/0902.1800

[11] Andreas Klappenecker and Martin Roetteler, *Engineering Functional Quantum Algorithms (2002)*, http://www.arxiv.org/abs/quant-ph/0208130

[12] E. Sejdić, I. Djurović, LJ. Stanković, "Fractional Fourier transform as a signal processing tool: An overview of recent developments," Signal Processing, vol. 91, no. 6, pp. 1351-1369, June 2011. doi:10.1016/j.sigpro.2010.10.008

[13] N. C. Pégard and J. W. Fleischer, "Optimizing holographic data storage using a fractional Fourier transform," Opt. Lett. 36, 2551-2553 (2011)

12.7   External links

- DiscreteTFDs -- software for computing the fractional Fourier transform and time-frequency distributions

- "Fractional Fourier Transform" by Enrique Zeleny, The Wolfram Demonstrations Project.

- Dr YangQuan Chen's FRFT (Fractional Fourier Transform) Webpages

- LTFAT - A free (GPL) Matlab / Octave toolbox Contains several version of the fractional Fourier transform.

12.8 Bibliography

- Ozaktas, Haldun M.; Zalevsky, Zeev; Kutay, M. Alper (2001), *The Fractional Fourier Transform with Applications in Optics and Signal Processing*, Series in Pure and Applied Optics, John Wiley & Sons, ISBN 0-471-96346-1

- Candan, C.; Kutay, M.A.; Ozaktas, H.M. (May 2000), "The discrete fractional Fourier transform", *IEEE Transactions on Signal Processing* **48** (5): 1329–1337, doi:10.1109/78.839980

- A. W. Lohmann, "Image rotation, Wigner rotation and the fractional Fourier transform," *J. Opt. Soc. Am.* **A 10**, 2181–2186 (1993).

- Soo-Chang Pei and Jian-Jiun Ding, "Relations between fractional operations and time-frequency distributions, and their applications," *IEEE Trans. Sig. Processing* **49** (8), 1638–1655 (2001).

- Jian-Jiun Ding, Time frequency analysis and wavelet transform class notes, the Department of Electrical Engineering, National Taiwan University (NTU), Taipei, Taiwan, 2007.

- Saxena, R., Singh, K., (2005) *Fractional Fourier transform: A novel tool for signal processing*, J. Indian Inst. Sci., Jan.–Feb. 2005, 85, 11–26. http://journal.library.iisc.ernet.in/vol200501/paper2/11.pdf.

Chapter 13

Non-uniform discrete Fourier transform

In applied mathematics, the **non-uniform discrete Fourier transform (NDFT)** of a signal is a type of Fourier transform, related to a discrete Fourier transform or discrete-time Fourier transform, but in which the input signal is not sampled at equally spaced intervals. As a result of this, the computed Discrete Fourier Transform can also consist of unevenly sampled frequency values. It is however also possible to compute uniformly sampled frequency values from an unevenly sampled input signal.

As a generalized approach for nonuniform sampling, NDFT can help us to get the information of a finite length signal in frequency domain at any frequency. It can also be used to design the FIR filters as the role of DFT, no matter if it's 1-D or 2-D.

One of the reasons to adopt NDFT is that most signals have their energy distributed nonuniformly in the frequency domain. Therefore, a nonuniform sampling scheme could be more convenient and useful in lots of applications of **Digital Signal Processing (DSP)**. For example, NDFT provides a variable spectral resolution controlled by the users.

Note that NDFT reduces to DFT when the sampling points are located on the unit circle at equally spaced angles.

13.1   1-D NDFT

13.1.1   Definition

1-D NDFT of a sequence x[n] of length N is[1]

$$X(z_k) = X(z)|_{z=z_k} = \sum_{n=0}^{N-1} x[n]z_k^{-n}, \quad k = 0, 1, ..., N-1,$$

where $X(z)$ is the Z-transform of $x[n]$, and $\{z_i\}_{i=0,1,...,N-1}$ are arbitrarily distinct points in the z-plane. Expressing the above as matrix, we get

$$\mathbf{X} = \mathbf{Dx}$$

where

$$\mathbf{X} = \begin{bmatrix} X(z_0) \\ X(z_1) \\ \vdots \\ X(z_{N-1}) \end{bmatrix}, \quad \mathbf{x} = \begin{bmatrix} x[0] \\ x[1] \\ \vdots \\ x[N-1] \end{bmatrix}, \text{and} \quad \mathbf{D} = \begin{bmatrix} 1 & z_0^{-1} & z_0^{-2} & \cdots & z_0^{-(N-1)} \\ 1 & z_1^{-1} & z_1^{-2} & \cdots & z_1^{-(N-1)} \\ \vdots & \vdots & \vdots & \ddots & \vdots \\ 1 & z_{N-1}^{-1} & z_{N-1}^{-2} & \cdots & z_{N-1}^{-(N-1)} \end{bmatrix}.$$

As we can see, the NDFT is characterized by $\mathbf{D}$ and hence the N z_k points. If we further factorize $det(\mathbf{D})$, we can see that $\mathbf{D}$ is nonsingular provided the N z_k points are distinct. If $\mathbf{D}$ is nonsingular, we can get a unique inverse NDFT as following:

$$\mathbf{x} = \mathbf{D}^{-1}\mathbf{X}$$

Given $\mathbf{X}$ and $\mathbf{D}$, we can use Gaussian elimination to solve $\mathbf{x}$. However, the complexity of this method is $O(N^3)$. To solve this problem more efficiently, we first determine $X(z)$ directly by polynomial interpolation

$$\hat{X}[k] = X(z_k), \quad k = 0, 1, ..., N-1,$$

then x[n] is the coefficients of the above interpolating polynomial which can be solved more efficiently. This is illustrated in the next subsection.

13.1.2   Solving The Inverse NDFT

Expressing $X(z)$ as the Lagrange polynomial of order N-1, we get

$$X(z) = \sum_{k=0}^{N-1} \frac{L_k(z)}{L_k(z_k)} \hat{X}[k],$$

where $\{L_i(z)\}_{i=0,1,...,N-1}$ are the fundamental polynomials:

$$L_k(z) = \prod_{i \neq k} (1 - z_i z^{-1}), \quad k = 0, 1, ..., N-1$$

Expressing $X(z)$ by Newton interpolation method, we get

$$X(z) = c_0 + c_1(1 - z_0 z^{-1}) + c_2(1 - z_0 z^{-1})(1 - z_1 z^{-1}) + ... + C_{N-1} \prod_{k=0}^{N-2} (1 - z_k z^{-1}),$$

where c_j is the divided difference of the jth order of $\hat{X}[0], \hat{X}[1], ..., \hat{X}[j]$ with respect to $z_0, z_1, ..., z_j$:

$$c_0 = \hat{X}[0],$$

$$c_1 = \frac{\hat{X}[1] - c_0}{1 - z_0 z_1^{-1}},$$

$$c_2 = \frac{\hat{X}[2] - c_0 - c_1(1 - z_0 z^{-1})}{(1 - z_0 z_2^{-1})(1 - z_1 z_2^{-1})},$$

$\vdots$

The disadvantage of Lagrange representation is that any additional point included will increase the order of the interpolating polynomial, leading to the need of recomputing all the fundamental polynomials. However, any additional point included in Newton representation only require one more term.

We can use a lower triangular system to solve $\{c_j\}$:

$$\mathbf{Lc} = \mathbf{X}$$

where

$$\mathbf{X} = \begin{bmatrix} \hat{X}[0] \\ \hat{X}[1] \\ \vdots \\ \hat{X}[N-1] \end{bmatrix}, \quad \mathbf{c} = \begin{bmatrix} c_0 \\ c_1 \\ \vdots \\ c_{N-1} \end{bmatrix}, \text{ and } \quad \mathbf{L}$$

$$= \begin{bmatrix} 1 & 0 & 0 & 0 & \cdots & 0 \\ 1 & (1 - z_0 z_1^{-1}) & 0 & 0 & \cdots & 0 \\ 1 & (1 - z_0 z_2^{-1}) & (1 - z_0 z_2^{-1})(1 - z_1 z_2^{-1}) & 0 & \cdots & 0 \\ \vdots & \vdots & \vdots & \vdots & \ddots & \vdots \\ 1 & (1 - z_0 z_{N-1}^{-1}) & (1 - z_0 z_{N-1}^{-1})(1 - z_1 z_{N-1}^{-1}) & \cdots & & \prod_{k=0}^{N-2}(1 - z_k z_{N-1}^{-1}) \end{bmatrix}.$$

By the above equation, $\{c_j\}$ can be computed within $O(N^3)$ operations. In this way Newton interpolation is more efficient than Lagrange Interpolation unless the latter is modified by

$$L_{k+1}(z) = \frac{(1 - z_{k+1} z^{-1})}{(1 - z_k z^{-1})} L_k(z), \quad k = 0, 1, ..., N - 1$$

13.2   2-D NDFT

2-D NDFT of a sequence $x[n_1, n_2]$ of size $N_1 \times N_2$ is[2]

$$\hat{X}(z_{1k}, z_{2k}) = \sum_{n_1=0}^{N_1-1} \sum_{n_2=0}^{N_2-1} x[n_1, n_2] z_{1k}^{-n_1} z_{2k}^{-n_2}, \quad k = 0, 1, ..., N_1 N_2 - 1,$$

where $\hat{X}(z_{1k}, z_{2k})$ is the 2-D z-transform of $x[n_1, n_2]$, and (z_{1k}, z_{2k}) are arbitrarily distinct $N_1 N_2$ points in the 4-D (z_1, z_2) space.

As in the 1-D case, we can express the above equation as

$$\hat{\mathbf{X}} = \mathbf{DX},$$

and the matrix $\mathbf{D}$ is also depends only on the choice of those sampling points. However, even if those sampling points are distinct, $\mathbf{D}$ could still be singular. No rules for determining whether the matrix is nonsingular or not have been found. Therefore, for all implementation of 2-D NDFT, we just check $det(\mathbf{D})$ for a specific set of sampling points. In general, the complexity of 2-D NDFT $O(N_1^3 N_2^3)$.

13.3   Applications

The applications of NDFT are:

- Digital filter design

- Spectral analysis

- Antenna array design

- Antenna pattern synthesis with prescribed nulls

- Decoding of dual-tone multi-frequency(DTMF) signals

- Dual-tone multi-frequency tone detection

13.4   See also

- Spectral estimation

13.5   External links

- Non-Uniform Fourier Transform: A Tutorial.

- "Nonuniform fast Fourier transforms using min-max interpolation". CiteSeerX: 10.1.1.15.3781.

- Notation, the NDFT and the NFFT

- NFFT 3.0 – Tutorial

13.6   Notes

[1] Marvasti 2001, p. 326.

[2] Marvasti 2001, p. 334.

13.7   References

- F. Marvasti, Nonuniform sampling: Theory and Practice. Plenum Publishers Co., 2001, pp. 325–360.

Chapter 14

Quantum Fourier transform

In quantum computing, the **quantum Fourier transform** is a linear transformation on quantum bits, and is the quantum analogue of the discrete Fourier transform. The quantum Fourier transform is a part of many quantum algorithms, notably Shor's algorithm for factoring and computing the discrete logarithm, the quantum phase estimation algorithm for estimating the eigenvalues of a unitary operator, and algorithms for the hidden subgroup problem.

The quantum Fourier transform can be performed efficiently on a quantum computer, with a particular decomposition into a product of simpler unitary matrices. Using a simple decomposition, the discrete Fourier transform on 2^n amplitudes can be implemented as a quantum circuit consisting of only $O(n^2)$ Hadamard gates and controlled phase shift gates, where n is the number of qubits.[1] This can be compared with the classical discrete Fourier transform, which takes $O(n2^n)$ gates (where n is the number of bits), which is exponentially more than $O(n^2)$. However, the quantum Fourier transform acts on a quantum state, whereas the classical Fourier transform acts on a vector, so not every task that uses the classical Fourier transform can take advantage of this exponential speedup.

The best quantum Fourier transform algorithms known today require only $O(n\log n)$gates to achieve an efficient approximation.[2]

14.1 Definition

The quantum Fourier transform is the classical discrete Fourier transform applied to the vector of amplitudes of a quantum state. The classical (unitary) Fourier transform acts on a vector in $\mathbb{C}^N$, $(x_0, ..., xN_{-1})$ and maps it to the vector $(y_0, ..., yN_{-1})$ according to the formula:

$$y_k = \frac{1}{\sqrt{N}} \sum_{j=0}^{N-1} x_j \omega^{jk}$$

where $\omega^{jk} = e^{2\pi i \frac{jk}{N}}$ is a Nth root of unity.

Similarly, the quantum Fourier transform acts on a quantum state $\sum_{i=0}^{N-1} x_i|i\rangle$ and maps it to a quantum state $\sum_{i=0}^{N-1} y_i|i\rangle$ according to the formula:

$$y_k = \frac{1}{\sqrt{N}} \sum_{j=0}^{N-1} x_j \omega^{jk},$$

with ω^{jk} defined as above.

This can also be expressed as the map

$$|j\rangle \mapsto \frac{1}{\sqrt{N}} \sum_{k=0}^{N-1} \omega^{jk}|k\rangle.$$

Equivalently, the quantum Fourier transform can be viewed as a unitary matrix (quantum gate, similar to a logic gate for classical computers) acting on quantum state vectors, where the unitary matrix F_N is given by

$$F_N = \frac{1}{\sqrt{N}} \begin{bmatrix} 1 & 1 & 1 & 1 & \cdots & 1 \\ 1 & \omega & \omega^2 & \omega^3 & \cdots & \omega^{N-1} \\ 1 & \omega^2 & \omega^4 & \omega^6 & \cdots & \omega^{2(N-1)} \\ 1 & \omega^3 & \omega^6 & \omega^9 & \cdots & \omega^{3(N-1)} \\ \vdots & \vdots & \vdots & \vdots & & \vdots \\ 1 & \omega^{N-1} & \omega^{2(N-1)} & \omega^{3(N-1)} & \cdots & \omega^{(N-1)(N-1)} \end{bmatrix}.$$

Here $\omega = e^{\frac{2\pi i}{N}}$ is a primitive N^{th} root of unity. For example, in the case of $N = 4$ we would find that $\omega = i$, so

$$F_4 = \frac{1}{2} \begin{bmatrix} 1 & 1 & 1 & 1 \\ 1 & i & -1 & -i \\ 1 & -1 & 1 & -1 \\ 1 & -i & -1 & i \end{bmatrix}.$$

14.2 Properties

14.2.1 Unitarity

Most of the properties of the quantum Fourier transform follow from the fact that it is a unitary transformation. This can be checked by performing matrix multiplication and ensuring that the relation $FF^\dagger = F^\dagger F = I$ holds, where $F^\dagger$ is the Hermitian adjoint of F. Alternately, one can check that vectors of norm 1 get mapped to vectors of norm 1.

From the unitary property it follows that the inverse of the quantum Fourier transform is the Hermitian adjoint of the Fourier matrix, thus $F^{-1} = F^\dagger$. Since there is an efficient quantum circuit implementing the quantum Fourier transform, the circuit can be run in reverse to perform the inverse quantum Fourier transform. Thus both transforms can be efficiently performed on a quantum computer.

14.3 Circuit implementation

The quantum Fourier transform can be approximately implemented for any N; however, the implementation for the case where N is a power of 2 is much simpler. Suppose $N = 2^n$. We have the orthonormal basis consisting of the vectors

$$|0\rangle, \ldots, |2^n - 1\rangle.$$

The basis states enumerate all the possible states of the qubits:

$$|x\rangle = |x_1 x_2 \ldots x_n\rangle = |x_1\rangle \otimes |x_2\rangle \otimes \cdots \otimes |x_n\rangle$$

where, with tensor product notation $\otimes$, $|x_j\rangle$ indicates that qubit j is in state x_j, with x_j either 0 or 1. By convention, the basis state index x orders the possible states of the qubits lexicographically, i.e., by converting from binary to decimal in this way:

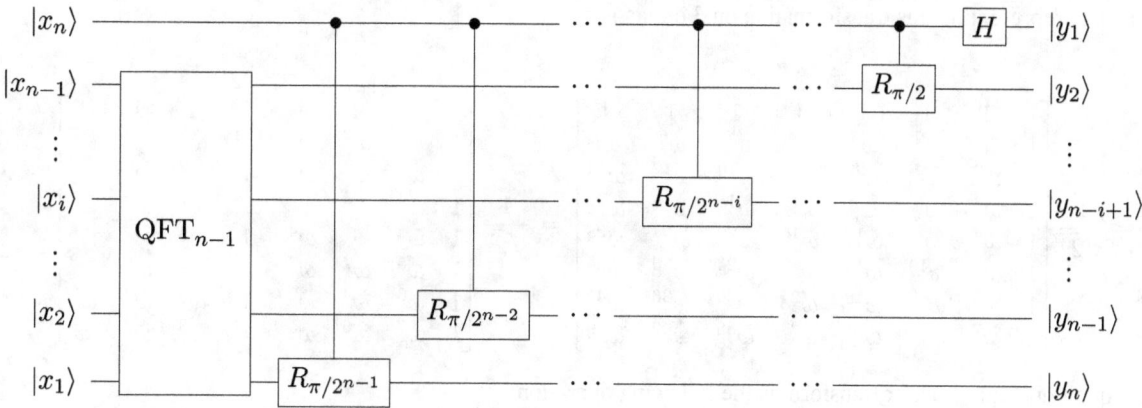

Quantum circuit representation of the quantum Fourier transform

$$x = x_1 2^{n-1} + x_2 2^{n-2} + \cdots + x_n 2^0.$$

It is also useful to borrow fractional binary notation:

$$[0.x_1 \ldots x_m] = \sum_{k=1}^{m} x_k 2^{-k}.$$

For instance, $[0.x_1] = \frac{x_1}{2}$ and $[0.x_1 x_2] = \frac{x_1}{2} + \frac{x_2}{2^2}$.

With this notation, the action of the quantum Fourier transform can be expressed as:

$$|x_1 x_2 \ldots x_n\rangle \mapsto \frac{1}{\sqrt{N}} \left(|0\rangle + e^{2\pi i [0.x_n]}|1\rangle \right) \otimes \left(|0\rangle + e^{2\pi i [0.x_{n-1} x_n]}|1\rangle \right) \otimes \cdots \otimes \left(|0\rangle + e^{2\pi i [0.x_1 x_2 \ldots x_n]}|1\rangle \right),$$

where the output qubit 1 is in a superposition of state $|0\rangle$ and $e^{2\pi i [0.x_n]}|1\rangle$, and so on for the other qubits.

In other words, the discrete Fourier transform, an operation on n-qubits, can be factored into the tensor product of n single-qubit operations, suggesting it is easily represented as a quantum circuit. In fact, each of those single-qubit operations can be implemented efficiently using a Hadamard gate and controlled phase gates. The first term requires one Hadamard gate, the next one requires a Hadamard gate and a controlled phase gate, and each following term requires an additional controlled phase gate. Summing up the number of gates gives $1 + 2 + \cdots + n = n(n+1)/2 = O(n^2)$ gates, which is polynomial in the number of qubits.

14.4 Example

Consider the quantum Fourier transform on 3 qubits. It is the following transformation:

$$|j\rangle \mapsto \frac{1}{\sqrt{2^3}} \sum_{k=0}^{2^3-1} \omega^{jk}|k\rangle,$$

where ω is a primitive eighth root of unity satisfying $\omega^8 = \left(e^{\frac{2\pi i}{8}} \right)^8 = 1$ (since $N = 2^3 = 8$).

The matrix representing this transformation on 3 qubits is

$$F_{2^3} = \frac{1}{\sqrt{2^3}} \begin{bmatrix} 1 & 1 & 1 & 1 & 1 & 1 & 1 & 1 \\ 1 & \omega & \omega^2 & \omega^3 & \omega^4 & \omega^5 & \omega^6 & \omega^7 \\ 1 & \omega^2 & \omega^4 & \omega^6 & \omega^8 & \omega^{10} & \omega^{12} & \omega^{14} \\ 1 & \omega^3 & \omega^6 & \omega^9 & \omega^{12} & \omega^{15} & \omega^{18} & \omega^{21} \\ 1 & \omega^4 & \omega^8 & \omega^{12} & \omega^{16} & \omega^{20} & \omega^{24} & \omega^{28} \\ 1 & \omega^5 & \omega^{10} & \omega^{15} & \omega^{20} & \omega^{25} & \omega^{30} & \omega^{35} \\ 1 & \omega^6 & \omega^{12} & \omega^{18} & \omega^{24} & \omega^{30} & \omega^{36} & \omega^{42} \\ 1 & \omega^7 & \omega^{14} & \omega^{21} & \omega^{28} & \omega^{35} & \omega^{42} & \omega^{49} \end{bmatrix} = \frac{1}{\sqrt{2^3}} \begin{bmatrix} 1 & 1 & 1 & 1 & 1 & 1 & 1 & 1 \\ 1 & \omega & \omega^2 & \omega^3 & \omega^4 & \omega^5 & \omega^6 & \omega^7 \\ 1 & \omega^2 & \omega^4 & \omega^6 & 1 & \omega^2 & \omega^4 & \omega^6 \\ 1 & \omega^3 & \omega^6 & \omega & \omega^4 & \omega^7 & \omega^2 & \omega^5 \\ 1 & \omega^4 & 1 & \omega^4 & 1 & \omega^4 & 1 & \omega^4 \\ 1 & \omega^5 & \omega^2 & \omega^7 & \omega^4 & \omega & \omega^6 & \omega^3 \\ 1 & \omega^6 & \omega^4 & \omega^2 & 1 & \omega^6 & \omega^4 & \omega^2 \\ 1 & \omega^7 & \omega^6 & \omega^5 & \omega^4 & \omega^3 & \omega^2 & \omega \end{bmatrix}.$$

The 3-qubit quantum Fourier transform is the following operation:

$$|x_1, x_2, x_3\rangle \mapsto \frac{1}{\sqrt{2^3}} \left(|0\rangle + e^{2\pi i [0.x_3]}|1\rangle \right) \otimes \left(|0\rangle + e^{2\pi i [0.x_2 x_3]}|1\rangle \right) \otimes \left(|0\rangle + e^{2\pi i [0.x_1 x_2 x_3]}|1\rangle \right).$$

This quantum circuit implements the quantum Fourier transform on the quantum state $|x_1, x_2, x_3\rangle$.

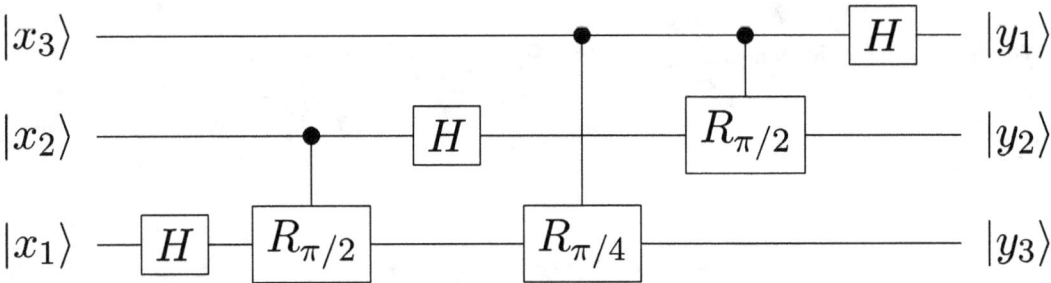

The quantum gates used in the circuit above are the Hadamard gate and the controlled phase gate R_θ.

As calculated above, the number of gates used is $n(n+1)/2$ which is equal to 6, for $n = 3$.

14.5 References

[1] Michael Nielsen and Isaac Chuang (2000). *Quantum Computation and Quantum Information*. Cambridge: Cambridge University Press. ISBN 0-521-63503-9. OCLC 174527496.

[2] L. Hales, S. Hallgren, An improved quantum Fourier transform algorithm and applications, Proceedings of the 41st Annual Symposium on Foundations of Computer Science, p. 515, November 12–14, 2000

- K. R. Parthasarathy, *Lectures on Quantum Computation and Quantum Error Correcting Codes* (Indian Statistical Institute, Delhi Center, June 2001)

- John Preskill, *Lecture Notes for Physics 229: Quantum Information and Computation* (CIT, September 1998)

14.6 External links

- Wolfram Demonstration Project: Quantum Circuit Implementing Grover's Search Algorithm

14.7 Text and image sources, contributors, and licenses

14.7.1 Text

Dav2008, Melchoir, Pgk, Jagged 85, Eskimbot, Skizzik, Oli Filth, Silly rabbit, Metacomet, Nbarth, Colonies Chris, Bob K, Tamfang, Dreadstar, Eliyak, Jim.belk, Lim Wei Quan, Dicklyon, Waggers, JoeBot, Martin Kozák, AlsatianRain, Paul Matthews, CRGreathouse, CmdrObot, Jackzhp, Shorespirit, HenningThielemann, Unmitigated Success, Myasuda, BigGoose2006, A.kverma, Kupirijo, Wrwrwr, ChrisKennedy, Entangledphotons, Doug Weller, Robertinventor, Hanche, Kablammo, The Hybrid, Jomoal99, Futurebird, Escarbot, BigJohnHenry, Sbandrews, Eleuther, Trlkly, Thenub314, Spamicles, Coffee2theorems, Richard Giuly, VoABot II, AuburnPilot, Hypergeek14, Jaakobou, JJ Harrison, 28421u2232nfenfcenc, Tenniszaz, MartinBot, Mårten Berglund, Nono64, JonathonReinhart, Lld2006, MistyMorn, Mike.lifeguard, Falquaddoomi, Jacksonwalters, Gombang, Policron, DorganBot, Juxtapos99, Natl1, LovaAndriamanjay, PowerWill500, Idioma-bot, VolkovBot, JohnBlackburne, LokiClock, AlnoktaBOT, Bovineboy2008, Philip Trueman, Beng186, One zero one, Hesam7, DennyColt, Mr. PIM, GirasoleDE, SieBot, Zbvhs, Faradayplank, RSStockdale, OKBot, Anchor Link Bot, Denisarona, Loren.wilton, ClueBot, Stokito, Mild Bill Hiccup, Excirial, Justin545, PixelBot, Brews ohare, Wikeepedian, Danielsimonjr, Ant59, DumZiBoT, AlanM1, XLinkBot, Galoisgroupie, Charles Sturm, Simplifix, RyanCross, Addbot, Olli Niemitalo, Fgnievinski, Chris19910, Fluffernutter, דולב, LaaknorBot, Ginosbot, Quercus solaris, Lightbot, Legobot, Publicly Visible, Luckas-bot, Yobot, The Earwig, Sarrus, SwisterTwister, Mlewko, AnomieBOT, Jim1138, Pete463251, Ehsfball78, Citation bot, Taeshadow, Andmats, PavelSolin, Obersachsebot, Xqbot, Bdmy, Adrian Wiemer, Jhbdel, RibotBOT, AliceNovak, Charithjayanada, Constructive editor, FrescoBot, Anterior1, Sławomir Biały, PiratePi, Citation bot 1, Lost-n-translation, Alipson, Jonesey95, Ashok567, Tcnuk, Rthimmig, TobeBot, Lotje, Afreiden, 777sms, Gzorg, Bj norge, Dalba, Kakahw, Fblasqueswiki, Newty23125, Teckcheong, EmausBot, KHamsun, Tawsifkhan, Dcirovic, Slawekb, Ὁ οἶστρος, SlimDeli, VishnuHaridas, A Thousand Doors, Maschen, Zueignung, Orange Suede Sofa, ChuispastonBot, Support.and.Defend, ClueBot NG, Jack Greenmaven, Dylan Moreland, Ilovejersey, Helpful Pixie Bot, Martin Berka, BG19bot, Walrus068, MadameBruxelles, Mark Arsten, Taylanmath, Glacialfox, BattyBot, Justincheng12345-bot, Jdogzz, JYBot, ZX95, Yashkes, Darvii, Federicofuentes, Cupitor, Tentinator, Pavel Bezdek, Friedlicherkoenig, Dpsangwal, 968u30u7, DominicPrice, Elenpach, Luocheng99, Loraof, Govkub, Philologick, Kavya l, KasparBot, Mikeyrichardson, 𝄞𝄢𝄢𝄢, Sebasgonky and Anonymous: 321

- **Generalized Fourier series** *Source:* https://en.wikipedia.org/wiki/Generalized_Fourier_series?oldid=607157858 *Contributors:* Michael Hardy, Charles Matthews, Stan Lioubomoudrov, Dysprosia, Giftlite, DemonThing, Oleg Alexandrov, Jftsang, MFH, Tawkerbot2, Nick Number, Rettetast, Brews ohare, Addbot, Zorrobot, Yobot, Erik9bot and Anonymous: 10

- **Discrete-time Fourier transform** *Source:* https://en.wikipedia.org/wiki/Discrete-time_Fourier_transform?oldid=679252999 *Contributors:* Michael Hardy, Ahoerstemeier, Stevenj, Furrykef, Giftlite, Dirk Gently, Discospinster, Rbj, LutzL, PAR, Cburnett, Oleg Alexandrov, Mwilde, Btyner, Fred Bradstadt, Alejo2083, Mathbot, YurikBot, RobotE, Axfangli, Zukeeper, Oli Filth, Metacomet, Nbarth, Darth Panda, Bob K, Ligulembot, Dicklyon, INKubusse, Goose240, Thijs!bot, Hazmat2, Frozenport, John254, Neilbartlett, Leyo, Krishnachandranvn, Anna Lincoln, SieBot, Jojalozzo, Kanonkas, Davyzhu, Wdwd, Addbot, AndersBot, Publicly Visible, Luckas-bot, Sz-iwbot, Citation bot, FrescoBot, Vaamarnath, RjwilmsiBot, JacobDang, GoingBatty, HiW-Bot, ZéroBot, Maschen, 28bot, Sitar Physics, ClueBot NG, Jack Greenmaven, Mattgately, Winner245, HMSSolent, BattyBot, Arcandam, Mark viking, Cupitor, Upputuri92, Monkbot and Anonymous: 74

- **Discrete Fourier transform** *Source:* https://en.wikipedia.org/wiki/Discrete_Fourier_transform?oldid=689202561 *Contributors:* AxelBoldt, Bryan Derksen, Zundark, Tbackstr, Gareth Owen, DrBob, Mjb, Edward, Michael Hardy, Booyabazooka, DopefishJustin, Loisel, SebastianHelm, Minesweeper, Mdebets, Cyp, Stevenj, Nikai, Charles Matthews, Dcoetzee, Dysprosia, Jitse Niesen, Matithyahu, Furrykef, Saltine, LMB, Omegatron, Ed g2s, Phil Boswell, Robbot, Astronautics~enwiki, MathMartin, Sverdrup, Wile E. Heresiarch, Tobias Bergemann, Connelly, Centrx, Giftlite, Ssd, Jorge Stolfi, Nayuki, Lockeownzj00, Bob.v.R, Glogger, Sam Hocevar, Kramer, Thorwald, ChrisRuvolo, Pak21, Paul August, Gauge, Alberto Orlandini, Iyerkri, Oyz, Foobaz, Minghong, Alexjs (usurped), PAR, Wtmitchell, Emvee~enwiki, Cburnett, Pgimeno~enwiki, Zxcvbnm, Gene Nygaard, Galaxiaad, Oleg Alexandrov, Kohtala, Woohookitty, Madmardigan53, Tabletop, Lmendo, Dzordzm, Cybedu~enwiki, Graham87, BD2412, Mulligatawny, Rjwilmsi, Crazyvas, HappyCamper, The wub, MikeJ9919, Mathbot, Srleffler, Chobot, YurikBot, Wavelength, Poslfit, RussBot, Epolk, Pseudomonas, David R. Ingham, Welsh, EverettColdwell, Ospalh, Hobit, Kompik, NormDor, Msuzen, Arthur Rubin, Teply, Bo Jacoby, Zvika, Sbyrnes321, Mlibby, SmackBot, Bluebot, Oli Filth, GregRM, Metacomet, Nbarth, DHN-bot~enwiki, Hongooi, Bob K, SheeEttin, Java410, Cybercobra, Borching~enwiki, Arialblack, Dicklyon, Kvng, MightyWarrior, Paul Matthews, CRGreathouse, CmdrObot, Hanspi, TheMightyOrb, HenningThielemann, Myasuda, Verdy p, Mcclurmc, Thijs!bot, Wikid77, Noel Bush, Marek69, DmitTrix, Sharpoo7, Escarbot, Fru1tbat, Mckee, Thenub314, Rabbanis, User A1, Martynas Patasius, Robin S, Jerome-Jerome, Bongomatic, Uncle Dick, Krishnachandranvn, Selinger, Epistemenical, Cuzkatzimhut, LokiClock, Nitin.mehta, Jwkuehne, Jobu0101, Hesam7, Felipebm, Swagato Barman Roy, AlleborgoBot, Peter Stalin, SieBot, CoolBlue1234, Flyer22 Reborn, Man It's So Loud In Here, Jdaloner, Svick, Anchor Link Bot, WikiBotas, Justin W Smith, Geoeg, Rdrk, Bill411432, Addbot, DOI bot, Fgnievinski, Ronhjones, Econo-Physicist, Uncia, Legobot, Luckas-bot, Yobot, OrgasGirl, Nallimbot, 1exec1, Пика Пика, Citation bot, Almabot, Herr Satz, AliceNovak, Samwb123, FrescoBot, Recognizance, Citation bot 1, Adlerbot, Davidmeo, Gryllida, Trappist the monk, DixonDBot, Jophos, Humatronic, Tumikk, David Binner, Bmitov, Helwr, EmausBot, John of Reading, Blin00, AsceticRose, Lorem Ip, Sampletalk, Maschen, Zueignung, ClueBot NG, Wbm1058, Bibcode Bot, DBigXray, Shakeel3442, Jeenriquez, ChrisGualtieri, Dexbot, Cerabot~enwiki, Citizentoad, Ben.k.horner, Acid1103, Benhorsburgh, Abevac, Brotherxandepuss, Haminoon, Fakufaku, Monkbot, IMochi, Linmanatee, Rewire91 and Anonymous: 222

- **List of Fourier-related transforms** *Source:* https://en.wikipedia.org/wiki/List_of_Fourier-related_transforms?oldid=654940093 *Contributors:* Toby Bartels, Rade Kutil, Michael Hardy, Nixdorf, Stevenj, Evercat, Charles Matthews, Maximus Rex, Omegatron, Robbot, Pdenapo, Giftlite, Gene Ward Smith, Seabhcan, Glogger, ZeroOne, Walden, Rbj, Forderud, Oleg Alexandrov, Linas, Andrei Polyanin, Ian Pitchford, Mathbot, YurikBot, Supten, Zzuuzz, Oli Filth, Bob K, Nmnogueira, Syrcatbot, TooMuchMath, Kvng, Ste4k, Thenub314, Parsecboy, Salih, OKBot, Wdwd, Loren.wilton, SchreiberBike, Addbot, OrgasGirl, Xqbot, Helwr, EmausBot, MaskedAce, Thambynayagam, Brad7777, Fuse809, Wstclyq, Nfvr and Anonymous: 14

- **Fast Fourier transform***Source:* https://en.wikipedia.org/wiki/Fast_Fourier_transform?oldid=690264505*Contributors:* AxelBoldt, The Anome,Tarquin, Gareth Owen, Roadrunner, DrBob, David spector, Boud, Michael Hardy, Pit~enwiki, Nixdorf, Pnm, Ixfd64, TakuyaMurata, Delir-ium, SebastianHelm, Stevenj, Palfrey, Hashar, Dcoetzee, Bemoeial, Dmsar, Jitse Niesen, Wik, Furrykef, Hyacinth, Grendelkhan, Cameronc,Bartosz, LMB, Vincent kraeutler, Twang, Donarreiskoffer, Robbot, Jaredwf, Fredrik, Academic Challenger, Yacht, Lupo, Giftlite, Dratman,Daniel Brockman, LiDaobing, Gunnar Larsson, Sam Hocevar, Mschlindwein, Adashiel, Guanabot, Smyth, Nagesh Anupindi, Ylai, Bender235,ZeroOne, Pt, Teorth, Jeltz, Artur Nowak~enwiki, Zxcvbnm, H2g2bob, Cxxl, Oleg Alexandrov, Mwilde, Decrease789, Eyreland, Eras-mus,Timendum, Marudubshinki, Qwertyus, Coneslayer, Rjwilmsi, Captain Disdain, Amitparikh, R.e.b., Fresheneesz, Kri, Chobot, YurikBot,Wavelength, David R. Ingham, Robertvan1, Herve661, Sangwine, CecilWard, Kkmurray, Squell, SmackBot, Geoffeg, Ulterior19802005, Uny-oyega, HalfShadow, TimBentley, Oli Filth, Nbarth, Zven, MaxSem, Audriusa, JustUser, LouScheffer, Spectrogram, DMPalmer, IronGargoyle,

16@r, JHunterJ, Rogerbrent, Norm mit, Domitori, MarylandArtLover, Lavaka, Vanished user sojweiorj34i4f, Requestion, HenningThiele-mann, Ntsimp, Djg2006, Sytelus, MauricioArayaPolo, Quibik, Headbomb, Electron9, LeoTrottier, BehnamFarid, Hcobb, Dawnseeker2000, Quintote, Gopimanne, Awilley, Thenub314, Timur lenk, Magioladitis, Hmo, JamesBWatson, Johnbibby, David Eppstein, MartinBot, Gah4, Crossz, Pankajp, Apexfreak, Adam Zivner, LokiClock, Omkar lon, Faestning, Felipebm, Maxim, Blablahblablah, ToePeu.bot, Le Pied-bot~enwiki, Anchor Link Bot, Melcombe, DonAByrd, Gene93k, GreenSpigot, Glutamin, Tuhinbhakta, Excirial, Tim32, Alexbot, Bender2k14, Rubin joseph 10, Arjayay, The Yowser, Qwfp, Efryevt, Herry41341, Dekart, Avalcarce, Hess88, Dioioib, Fgnievinski, 2ganesh, MrOllie, Lightbot, EugeneZ, Apteva, Legobot, Luckas-bot, Yobot, Twexcom, Akoesters, QueenCake, AnomieBOT, Jim1138, Materialscientist, Dead-Totoro, ArthurBot, Xqbot, Minibikini, David A se, AliceNovak, Smallman12q, Steina4, Aclassifier, FrescoBot, Klokbaske, Carbone1853, Citation bot 1, Boxplot, Wikichicheng, Dcwill1285, Davidmeo, Kellybundybrain, Gryllida, Bmitov, Jfmantis, Solongmarriane, Helwr, K6ka, R. J. Mathar, Dondervogel 2, Staszek Lem, Kuashio, Lorem Ip, Maschen, Cgt, Riemann'sZeta, Plantdrew, Koertefa, Stelpa, SciCompTeacher, Klilidiplomus, Haynals, Eflatmajor7th, YFdyh-bot, Sandeep.ps4, Donn300, Harsh 2580, Kushalsatrasala, Ogmark, FFTguy, Junkyardsparkle, Icarot, Michipedian, Debouch, Sattar91, LCS check, Mathtruth, BioFluid, Iustin Diac, Monkbot, ErRied, D1ofBerks, ChiCheng, 115ash, Oiyarbepsy, Nir.ailon, Thesejma, 89sec, PurpleLego, Draak13, Shaddowffax and Anonymous: 212

- **Time–frequency analysis** *Source:* https://en.wikipedia.org/wiki/Time%E2%80%93frequency_analysis?oldid=675671437 *Contributors:* MichaelHardy, Charles Matthews, Peter M Gerdes, Daniel Mietchen, KasugaHuang, SmackBot, PEHowland, Nbarth, OrphanBot, Just plain Bill,Dicklyon, Lavaka, Alaibot, Conquerist, Typometer, Cuzkatzimhut, Jesin, Cwkmail, Melcombe, Aquegg, NTUDISP, EelkeSpaak, Lautaro2k,MilesTerrex, Addbot, Fgnievinski, Yobot, Legobot II, Torresol, Erik9bot, John of Reading, GoingBatty, Nasseroleslami, Boashash, Burhem,Shwab, Poolpeggy, Jamshid.parsi, Helpful Pixie Bot, NotWith, Jewel406 and Anonymous: 21

- **Short-time Fourier transform** *Source:* https://en.wikipedia.org/wiki/Short-time_Fourier_transform?oldid=685493563 *Contributors:* Michael Hardy, Dcljr, Stevenj, Charles Matthews, Furrykef, Omegatron, Chris Roy, Giftlite, BenFrantzDale, Rich Farmbrough, FafnerAzugon, STHay-den, Root4(one), Rbj, Johnteslade, Rush3k, Atlant, Keenan Pepper, PAR, Alejo2083, Mathbot, YurikBot, Manop, Gaius Cornelius, 48v, Ka-sugaHuang, Edin1, SmackBot, Reedy, Gilliam, Oli Filth, Jdh30, Nbarth, Dicklyon, Kvng, Tawkerbot2, Thijs!bot, D4g0thur, Nick Number, DFTDER~enwiki, David Eppstein, Glrx, TXiKiBoT, Pantelis vassilakis, Aquegg, PipepBot, Mild Bill Hiccup, PixelBot, Arjayay, Steven-cys, Dkondras, MystBot, Addbot, Fgnievinski, Tide rolls, Luckas-bot, Yobot, Gabriele Nunzio Tornetta, AnomieBOT, CBMalloch, Rubinbot, Klokbaske, RedBot, Nellatnoj, DixonDBot, Some Wiki Editor, Ripchip Bot, EmausBot, John of Reading, WikitanvirBot, Slawekb, ZéroBot, TarryWorst, Boashash, ClueBot NG, Vacation9, Shantham11, Martin Berka, Luegge, Bjfwiki, A4b3c2d1e0f, Babitaarora, Monkbot, Psoen-derg, Tema.emelyan, Fourier1789 and Anonymous: 55

- **Gabor transform** *Source:* https://en.wikipedia.org/wiki/Gabor_transform?oldid=684589365 *Contributors:* Michael Hardy, Giftlite, BD2412, SmackBot, Reza mirhosseini, David Eppstein, Quietbritishjim, JerroldPease-Atlanta, Stevencys, UnCatBot, Kolyma, Addbot, AnomieBOT, BillHart93, BG19bot, Eflatmajor7th, Wiwi Samsul, Imhotaru and Anonymous: 17

- **Fractional Fourier transform** *Source:* https://en.wikipedia.org/wiki/Fractional_Fourier_transform?oldid=675531629 *Contributors:* Michael Hardy, Stevenj, AllanR~enwiki, Charles Matthews, Giftlite, Glogger, Chadernook, Billlion, PAR, Linas, Teply, KasugaHuang, SmackBot, Nbarth, Jhealy, Md2perpe, TomyDuby, Cuzkatzimhut, SieBot, Sabri76, Stevencys, XLinkBot, Marraco~enwiki, Addbot, Ezelenyv, Light-bot, Yobot, Citation bot 1, EmausBot, Johngambot, Boashash, Sitar Physics, Helpful Pixie Bot, BG19bot, Rhodan21, Gui23452, BattyBot, Grantjune, ChrisGualtieri, Wanttoknow, Monkbot, BALTAM Itatsu, Psoenderg, Gholleywiki and Anonymous: 20

- **Non-uniform discrete Fourier transform** *Source:* https://en.wikipedia.org/wiki/Non-uniform_discrete_Fourier_transform?oldid=640406040 *Contributors:* Michael Hardy, Kku, Stevenj, Phil Boswell, Joriki, Chris the speller, Sun Creator, Avalcarce, Fgnievinski, Dannyboytward, Per Ardua, Ytw1987, Cameronroytaylor and Anonymous: 1

- **Quantum Fourier transform** *Source:* https://en.wikipedia.org/wiki/Quantum_Fourier_transform?oldid=677165733 *Contributors:* Michael Hardy, Charles Matthews, MathMartin, DemonThing, CSTAR, Creidieki, Sardanixka, Squizzz~enwiki, KasugaHuang, Gilliam, Lamarth, CapitalR, Petr Matas, Mct mht, David Eppstein, Yonidebot, Pyrospirit, Squids and Chips, LokiClock, Jesin, Quantquark, Raydenlighter, Ben-der2k14, DumZiBoT, XLinkBot, Addbot, PaulTheBaker, Yobot, RobinK, EmausBot, Kodus, Elprofediaz, Syberspot and Anonymous: 18

14.7.2   Images

- **File:AtomicOrbital_n4_l2.png** *Source:* https://upload.wikimedia.org/wikipedia/commons/0/04/AtomicOrbital_n4_l2.png *License:* CC-BY-SA-3.0 *Contributors:* Transferred from en.wikipedia to Commons by Zinder using CommonsHelper. *Original artist:* The original uploader was DMacks at English Wikipedia

- **File:Bass_Guitar_Time_Signal_of_open_string_A_note_(55_Hz).png** *Source:* https://upload.wikimedia.org/wikipedia/commons/a/a6/Bass_Guitar_Time_Signal_of_open_string_A_note_%2855_Hz%29.png *License:* CC BY-SA4.0 *Contributors:* Own work *Original artist:* Fourier1789

- **File:Bloch_Sphere.svg** *Source:* https://upload.wikimedia.org/wikipedia/commons/f/f4/Bloch_Sphere.svg *License:* CC BY-SA 3.0 *Contributors:* Own work *Original artist:* Glosser.ca

- **File:Commons-logo.svg** *Source:* https://upload.wikimedia.org/wikipedia/en/4/4a/Commons-logo.svg *License:* ? *Contributors:* ? *Original artist:* ?

- **File:Commutative_diagram_illustrating_problem_solving_via_the_Fourier_transform.svg** *Source:* https://upload.wikimedia.org/wikipedia/commons/f/fd/Commutative_diagram_illustrating_problem_solving_via_the_Fourier_transform.svg *License:* CC BY-SA 3.0 *Contributors:* Own work *Original artist:* Quietbritishjim

- **File:Filter_fractional.jpg** *Source:* https://upload.wikimedia.org/wikipedia/commons/5/59/Filter_fractional.jpg *License:* Public domain *Contributors:* Transferred from en.wikipedia to Commons. *Original artist:* NTUDISP at English Wikipedia

- **File:Filter_tf.jpg** *Source:* https://upload.wikimedia.org/wikipedia/commons/7/72/Filter_tf.jpg *License:* Public domain *Contributors:* Transferred from en.wikipedia to Commons. *Original artist:* NTUDISP at English Wikipedia

- **File:Folder_Hexagonal_Icon.svg** *Source:* https://upload.wikimedia.org/wikipedia/en/4/48/Folder_Hexagonal_Icon.svg *License:* Cc-by-sa-3.0 *Contributors:* ? *Original artist:* ?

14.7.3 Content license

www.ingramcontent.com/pod-product-compliance
Lightning Source LLC
Chambersburg PA
CBHW081149180526
45170CB00006B/2002